Local Approximation of the Holstein Polaron Problem

Jean-Marc ROBIN

Contents

1 Introduction

The Holstein model [1] describes non interacting electrons on a lattice coupled to dispersionless phonons. For one electron in the system, the bare electron is dressed by a cloud of bosonic excitations and is called a polaron. The properties of this new particle are still puzzling, although this problem is very old and has been solved numerically for finite or infinite size systems [2]. Most of these works are devoted to study the Green's function of the bare electron, $G(\mathbf{k}, \omega)$, which can be measured by photoemission experiments, and not to understand the fundamental nature of the polaron as a dressed particle [3, 4]. The difficulty arise from the fact that only in the atomic limit of the problem, a small polaron operator can be mathematically constructed, and its Green's function $\tilde{G}(\omega)$ defined.

Within the Local Impurity Self-Consistent Approximation [5], which becomes exact in the limit of infinite dimensions, the analytical expression of the local Green's function for the electron [6] has strong simililarities with its counterpart in the atomic limit, so it is worth to review this local approximation in the spirit of the small polaron physics, and to learn how to compute the Green's function $\tilde{G}(\mathbf{k}, z)$ of the small polaron.

For one electron in the system, the Hamiltonian reads

$$H = \varepsilon_0 \sum_j c_j^\dagger c_j - t_0 \sum_{j,\delta} c_{j+\delta}^\dagger c_j + \omega_0 \sum_j b_j^\dagger b_j \tag{1}$$

$$- g_0 \omega_0 \sum_j c_j^\dagger c_j (b_j^\dagger + b_j),$$

where $c_j^\dagger$ and c_j creates and annihilates an electron at site $\mathbf{R}_j$, $b_j^\dagger$ and b_j creates and annihilates a bosonic excitation at site $\mathbf{R}_j$. The sum over j runs over the M sites of the lattice, the sum over δ runs over nearest neighbors, ω_0 is the optical frequency of the phonons, ε_0 is the atomic energy, t_0 is the hopping energy, and g_0 is a dimensionless coupling constant. We define the basis states of the phonons as $|\mathbf{n}\rangle$, with $b_j^\dagger b_j |\mathbf{n}\rangle = n_j |\mathbf{n}\rangle$. For one electron in the system, the basis states are $c_j^\dagger |\mathbf{n}\rangle$. At zero temperature, the polaron problem is to compute the Green's function for the electron

$$G_{i,j}^{\mathbf{m},\mathbf{n}}(z) = \langle \mathbf{m} | c_i \frac{1}{z - H} c_j^\dagger | \mathbf{n} \rangle \tag{2}$$

This Green's function contains all the physics of the problem. On the ground state of the phonons, that is $G_{i,j}^{0,0}(z)$, we have just access to the excitations of the bare electron, $G(\mathbf{k},\omega)$. In this paper, we focus on the Green's function of the small polaron,

$$\tilde{G}_{i,j}^{0,0}(z) = \langle \mathbf{0}|U_i^\dagger c_i \frac{1}{z - H} c_j^\dagger U_j|\mathbf{0}\rangle \tag{3}$$

where U_j is a local operator defined from the atomic limit of the problem.

2   Basic ideas of DMFT

The Dynamical Mean Field Theory, or DMFT, deals with correlated electrons on a lattice and reduces the problem to a single site embedded self-consistently in an effective medium [5], so the the first idea is to focus on a single site, o,

$$H = H^{(o)} - t_0 \sum_\delta c_{o+\delta}^\dagger c_o + H^{(\odot)} \tag{4}$$

with

$$H^{(o)} = \varepsilon_0 c_o^\dagger c_o + \omega_0 b_o^\dagger b_o - g_0\omega_0(b_o^\dagger + b_o)c_o^\dagger c_o \tag{5}$$

and to define an effective Hamiltonian for this single site

$$H_o = H^{(o)} + H_W^{(o)}. \tag{6}$$

The term $H_W^{(o)}$ describes the effective medium.

Consider the one particle Green's function for the electron at this site o on the vacuum of the phonons

$$G_{o,o}(t) = \frac{\theta(t)}{i\hbar} \langle \mathbf{0}|c_o(t)c_o^\dagger|\mathbf{0}\rangle. \tag{7}$$

We obtain the equation of evolution

$$i\hbar\partial_t G_{o,o}(t) = \delta(t) + \varepsilon_0 G_{o,o}(t) - t_0 \sum_\delta G_{o+\delta,o}(t) + F_{o,o}(t) \tag{8}$$

with

$$F_{o,o}(t) = \frac{\theta(t)}{i\hbar} \langle \mathbf{0}|[c_o, H^{(o)} - \varepsilon_0 c_o^\dagger c_o](t)c_o^\dagger|\mathbf{0}\rangle. \tag{9}$$

Going to complex valued frequencies, we get

$$zG_{o,o}(z) = 1 + \varepsilon_0 G_{o,o}(z) - t_0 \sum_\delta G_{o+\delta,o}(z) + F_{o,o}(z). \tag{10}$$

We see that we can obtain a closed solution for the local Green's function if we can define a local Weiss's self-energy

$$-t_0 \sum_\delta G_{o+\delta,o}(z) = W_{o,o}(z)G_{o,o}(z) \tag{11}$$

and a local interaction self-energy

$$F_{o,o}(z) = \Sigma_{o,o}(z)G_{o,o}(z). \tag{12}$$

Then the solution is

$$G_{o,o}(z) = \frac{1}{z - \varepsilon_0 - W_{o,o}(z) - \Sigma_{o,o}(z)}. \tag{13}$$

Now, without approximation, for a periodic lattice, we know that the equation of evolution has the form

$$zG(\mathbf{k}, z) = 1 + \varepsilon_0 G(\mathbf{k}, z) + \xi_k G(\mathbf{k}, z) + \Sigma(\mathbf{k}, z)G(\mathbf{k}, z). \tag{14}$$

We have introduced the Fourier transforms

$$c_j = \frac{1}{\sqrt{M}} \sum_\mathbf{k} e^{i\mathbf{k}\mathbf{R}_j} c_\mathbf{k}, \tag{15}$$

$$G_{i,j}(z) = \frac{1}{M} \sum_\mathbf{k} e^{i\mathbf{k}(\mathbf{R}_i - \mathbf{R}_j)} G(\mathbf{k}, z), \tag{16}$$

$$-t_0 \sum_\delta G_{i+\delta,i}(z) = \sum_\mathbf{k} \xi_\mathbf{k} G(\mathbf{k}, z). \tag{17}$$

If we choose $\Sigma(\mathbf{k}, z) = \Sigma(z)$, we obtain

$$\frac{1}{M} \sum_\mathbf{k} \xi_\mathbf{k} G(\mathbf{k}, z) = W(z)G(z) \tag{18}$$

with

$$G(\mathbf{k}, z) = \frac{1}{z - \varepsilon_0 - \xi_\mathbf{k} - \Sigma(z)} \tag{19}$$

and

$$G(z) \;=\; \frac{1}{M}\sum_{\mathbf{k}} G(\mathbf{k},z) \;=\; \frac{1}{z - \varepsilon_0 - W(z) - \Sigma(z)}. \tag{20}$$

We thus see that we can set up a local approximation for our lattice Hamiltonian if we put by hand that the self-energy is local, and get a Weiss's self-energy that depends only of the lattice and of this local self-energy.

Let us consider a two-site problem. In this case we have $\xi_{\mathbf{k}} = \pm t_0$, so the local Green's function is given by

$$G(z) = \frac{1}{2}\left[\frac{1}{z - \varepsilon_0 - t_0 - \Sigma(z)} + \frac{1}{z - \varepsilon_0 + t_0 - \Sigma(z)}\right], \tag{21}$$

and we obtain the expression of the Weiss's self-energy

$$W(z) \;=\; \frac{t_0^2}{z - \varepsilon_0 - \Sigma(z)}. \tag{22}$$

At this stage, we introduce the Renormalized Perturbation Expansion, or RPE [7]. Within the DMFT, the equation of evolution of the Green's function, for a periodic lattice, is

$$z G_{i,j}(z) = \delta_{i,j} + \varepsilon_0 G_{i,j}(z) - t_0 \sum_{\delta} G_{i+\delta,j}(z) + \Sigma(z) G_{i,j}(z). \tag{23}$$

We can thus define a free propagator

$$G_{i,j}^{(0)}(z) \;=\; \frac{\delta_{i,j}}{z - \varepsilon_0 - \Sigma(z)} \tag{24}$$

and set a Dyson's equation with the hopping term as the non diagonal part of the Hamiltonian. Figure 1(a) corresponds to the contribution

$$G_{i,j} \;=\; G_{i,i}^{(0)} \, t_0 G_{n1,n1}^{(0)} \, t_0 G_{j,j}^{(0)} \tag{25}$$

and Figure 1(b) corresponds to the contribution

$$G_{i,i} \;=\; G_{i,i}^{(0)} \, t_0 G_{n1,n1}^{(0)} \, t_0 G_{n2,n2}^{(0)} \, t_0 G_{n3,n3}^{(0)} \, t_0 G_{i,i}^{(0)}. \tag{26}$$

The RPE states that one uses renormalized Green's functions as we sum over all paths. The above contribution for $G_{i,j}$ becomes

$$G_{i,j} \;=\; G_{i,i} \, t_0 G_{n1,n1[i]} \, t_0 G_{j,j,[i,n1]} \tag{27}$$

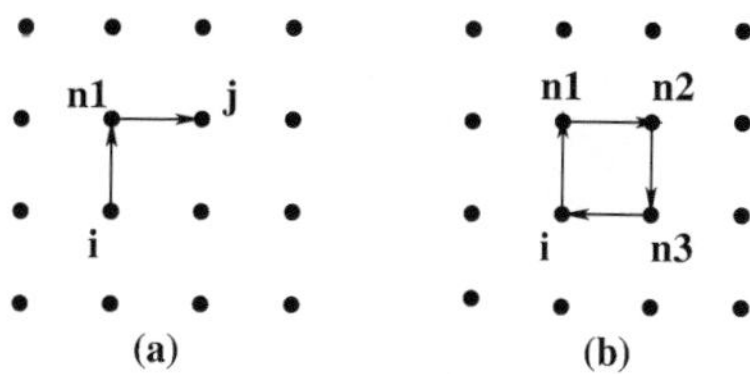

Figure 1: Path (a) is a contribution to $G_{i,j}$ and path (b) to $G_{i,i}$.

where $G_{n1,n1[i]}$ is the renormalized Green's function for site $n1$ for a lattice where the site i has been removed, and $G_{j,j,[i,n1]}$ is the renormalized Green's function for site j where sites i and $n1$ have been removed from the lattice. The contribution for $G_{i,i}$ becomes

$$G_{i,i} \;=\; G_{i,i}\, t_0 G_{n1,n1[i]}\, t_0 G_{n2,n2[i,n1]}\, t_0 G_{n3,n3[i,n1,n2]}\, t_0 G_{i,i}^{(0)}. \tag{28}$$

The last factor $G_{i,i}^{(0)}$ is the unrenormalized Green's function since the first factor $G_{i,i}$ takes account of the renormalization.

Let us consider a one dimensional infinite lattice. The equation for $G_{i,i}$ is

$$G_{i,i} = G_{i,i}^{(0)} + t_0^2 G_{i,i} G_{i+1,i+1[i]} G_{i,i}^{(0)} + t_0^2 G_{i,i} G_{i-1,i-1[i]} G_{i,i}^{(0)}. \tag{29}$$

Within the RPE, the path $i \to i+1 \to i+2 \to i+1 \to i$ is not allowed since the site $i + 1$ has been removed of the lattice. Next, we get

$$G_{i+1,i+1[i]} = G_{i+1,i+1}^{(0)} + t_0^2 G_{i+1,i+1[i]} G_{i+2,i+2[i,i+1]} G_{i+1,i+1}^{(0)}. \tag{30}$$

We notice that $G_{i+2,i+2[i,i+1]} = G_{i+2,i+2[i+1]}$. We thus obtain the solution for the local Green's function as a continued fraction expansion,

$$G_{i,i}(z) \;=\; \frac{1}{z - \varepsilon_0 - \Sigma(z) - 2t_0^2 \Phi(z)} \tag{31}$$

with the self-consistent equation for $\Phi(z)$,

$$\Phi(z) \;=\; \frac{1}{z - \varepsilon_0 - \Sigma(z) - t_0^2 \Phi(z)}. \tag{32}$$

For the two-site cluster, we obtain the equations

$$G_{1,1} = G_{1,1}^{(0)} + t_0^2 G_{1,1} G_{2,2[1]} G_{1,1}^{(0)} \tag{33}$$

and $G_{2,2[1]} = G_{2,2}^{(0)}$. The local Green's function is thus

$$G_{1,1}(z) = \cfrac{1}{z - \varepsilon_0 - \Sigma(z) - \cfrac{t_0^2}{z - \varepsilon_0 - \Sigma(z)}}. \tag{34}$$

For the Bethe lattice with Z neighbors, we obtain

$$G_{i,i} = G_{i,i}^{(0)} + \sum_\delta t_0^2 G_{i,i} G_{i+\delta,i+\delta[i]} G_{i,i}^{(0)} \tag{35}$$

and

$$G_{i+\delta,i+\delta[i]} = G_{i+\delta,i+\delta}^{(0)} \tag{36}$$

$$+ \sum_{\delta'} t_0^2 G_{i+\delta,i+\delta[i]} G_{i+\delta+\delta',i+\delta+\delta'[i,i+\delta]} G_{i+\delta,i+\delta}^{(0)}.$$

For the site i there is Z indentical contributions, while for site $i + \delta$ there is only $Z - 1$ contributions since the path back to i is forbiden. We obtain the solution for the local Green's function

$$G_{i,i}(z) = \frac{1}{z - \varepsilon_0 - \Sigma(z) - Z t_0^2 \Phi(z)} \tag{37}$$

with the recursive expansion

$$\Phi(z) = \frac{1}{z - \varepsilon_0 - \Sigma(z) - (Z - 1) t_0^2 \Phi(z)}. \tag{38}$$

Now, if we take the limit $Z \to \infty$ with the scaling $t_0 \to t_0/\sqrt{Z}$, we obtain $W(z) = t_0^2 G(z)$.

Consider now a non periodic lattice. In this case the equations of evolution, within the DMFT, are

$$z G_{i,j}(z) = \delta_{i,j} + \varepsilon_0 G_{i,j}(z) + \Sigma_i(z) G_{i,j}(z) - t_0 \sum_\delta G_{i+\delta,j}(z) \tag{39}$$

and we define the free propagators of the RPE as

$$G_{i,i}^{(0)}(z) = \frac{1}{z - \varepsilon_0 - \Sigma_i(z)}. \tag{40}$$

For a three-site problem, with open boundary conditions, we get the equations of evolution

$$\begin{cases} zG_{1,1}(z) = 1 + (\varepsilon_0 + \Sigma_1(z))G_{1,1}(z) - t_0 G_{1,2}(z) \\[2mm] zG_{2,1}(z) = (\varepsilon_0 + \Sigma_2(z))G_{2,1}(z) - t_0 G_{1,1}(z) - t_0 G_{3,1}(z) \\[2mm] zG_{3,1}(z) = (\varepsilon_0 + \Sigma_3(z))G_{3,1}(z) - t_0 G_{2,1}(z) \end{cases} \tag{41}$$

In the language of the RPE, we write

$$\begin{cases} G_{1,1}(z) & = & G_{1,1}^{(0)}(z) + t_0^2 G_{1,1}(z) G_{2,2[1]}(z) G_{1,1}^{(0)}(z) \\[2mm] G_{2,2[1]}(z) & = & G_{2,2}^{(0)}(z) + t_0^2 G_{2,2[1]}(z) G_{3,3[1,2]}(z) G_{2,2}^{(0)}(z) \\[2mm] G_{3,3[1,2]}(z) & = & G_{3,3}^{(0)}(z) \end{cases} \tag{42}$$

and obtain the solution as

$$G_{1,1}(z) = \frac{1}{z - \varepsilon_0 - \Sigma_1(z) - W_1(z)} \tag{43}$$

with

$$W_1(z) = \frac{t_0^2}{z - \varepsilon_0 - \Sigma_2(z) - \dfrac{t_0^2}{z - \varepsilon_0 - \Sigma_3(z)}}. \tag{44}$$

For $G_{2,2}(z)$, we obtain

$$G_{2,2}(z) = \frac{1}{z - \varepsilon_0 - \Sigma_2(z) - W_2(z)} \tag{45}$$

with

$$W_2(z) = \frac{t_0^2}{z - \varepsilon_0 - \Sigma_1(z)} + \frac{t_0^2}{z - \varepsilon_0 - \Sigma_3(z)}. \tag{46}$$

In this case, $\Sigma_1(z) = \Sigma_3(z)$ and $W_1(z) = W_3(z)$. We notice that $G_{1,2}(z) = -t_0 G_{1,1}(z) G_{2,2[1]}(z)$ and $G_{1,3}(z) = t_0^2 G_{1,1}(z) G_{2,2[1]}(z) G_{3,3[1,2]}(z)$, i.e. that all the non-diagonal Green's functions can be expressed in terms of local Green's functions (that may differ by their Weiss's self-energies). We notice, also, that the Weiss's self-energy of a given site involves only the interacting self-energies of the other sites.

3 Atomic Solution

For a single site, the Hamiltonian reads

$$H_{at} = \varepsilon_0 c^\dagger c + \omega_0 b^\dagger b - g_0 \omega_0 c^\dagger c (b^\dagger + b). \tag{47}$$

The basis states of the phonons are $|n\rangle$, with $b^\dagger b |n\rangle = n|n\rangle$. The Hamiltonian commutes with the number of electron $c^\dagger c$, so one can diagonalize it in each subspace. For $c^\dagger c = 0$, one gets an harmonic oscillator with $H_{at}|n\rangle = n\omega_0|n\rangle$. For $c^\dagger c = 1$, one gets a displaced harmonic oscillator with $H_{at}|n) = (\varepsilon_0 - g_0^2 \omega_0 + n\omega_0)|n)$. The eigenstates are $|n) = U|n\rangle$ where U is the unitary operator

$$U = e^{g_0(b^\dagger - b)} = e^{-\frac{1}{2}g_0^2} e^{g_0 b^\dagger} e^{-g_0 b} \tag{48}$$

with the properties $U^\dagger b U = b + g_0$ and $U^\dagger b^\dagger U = b^\dagger + g_0$ [8]. The Green's function of the electron is given by

$$G_{at}^{m,n}(z) = \langle m|c \frac{1}{z - H_{at}} c^\dagger |n\rangle. \tag{49}$$

Since we compute it for one electron, we get

$$G_{at}^{m,n}(z) = \langle m| \frac{1}{z - \varepsilon_0 - \omega_0 b^\dagger b + g_0 \omega_0 (b^\dagger + b)} |n\rangle. \tag{50}$$

Next, we define the atomic polaron, or small polaron, creation operator as $\tilde{c}^\dagger = U c^\dagger$. When acting on the state $|n\rangle$ it creates an electron and a coherent state $|n) = U|n\rangle$. The Green's function of the small polaron is given by

$$\tilde{G}_{at}^{m,n}(z) = \langle m|U^\dagger c \frac{1}{z - H_{at}} c^\dagger U |n\rangle = \frac{\delta_{m,n}}{z - \varepsilon_0 + g_0^2 \omega_0 - m\omega_0}. \tag{51}$$

8

Instead of writing matrix elements, one can directly write the operator form of these local Green's functions:

$$\mathbf{G}_{at}(z) = \frac{1}{z - \varepsilon_0 - \omega_0 b^\dagger b + g_0 \omega_0 X} \tag{52}$$

and

$$\tilde{\mathbf{G}}_{at}(z) = \frac{1}{z - \tilde{\varepsilon}_0 - \omega_0 b^\dagger b} \tag{53}$$

with $\tilde{\varepsilon}_0 = \varepsilon_0 - g_0^2 \omega_0$ and $X = b^\dagger + b$. We notice the relation $\tilde{\mathbf{G}}_{at}(z) = U^\dagger \mathbf{G}_{at}(z) U$.

Finally, we introduce the unitary transformation that diagonalize H_{at}, such that $e^S = 1$ for $c^\dagger c = 0$ and $e^S = U$ for $c^\dagger c = 1$. This is just $S = g_0 c^\dagger c(b^\dagger - b)$. One obtains the transformed Hamiltonian as

$$\tilde{H}_{at} = e^{-S} H e^S = \tilde{\varepsilon}_0 c^\dagger c + \omega_0 b^\dagger b. \tag{54}$$

It describes a free fermionic small polaron and free bosonic excitations.

4 Standard DMFT solution

In the Local Impurity Self-Consistent Approximation [5], one defines a local Hamiltonian H_{loc} for an given site o, which contains the atomic Hamiltonian H_{at} and a Weiss's Hamiltonian H_W, given by

$$H_W = \sum_k \varepsilon_k c_k^\dagger c_k + \sum_k v_k [c_k^\dagger c + c^\dagger c_k]. \tag{55}$$

Here, c_k and $c_k^\dagger$ are auxiliary fermionic annihilation and creation operators. They mimic the contribution of the other sites of the lattice. The parameters ε_k and v_k are chosen to fit the Weiss's self-energy of this given site, $W(z)$. Since there is only one electron in our problem, we have the condition $c^\dagger c + \sum_k c_k^\dagger c_k = 1$. The basis states of the problem are $c^\dagger |n\rangle$ and $c_k^\dagger |n\rangle$, so we can define Green's functions for these auxiliary fermions as

$$G_{k,k}^{m,n}(z) = \langle m | c_k \frac{1}{z - H_{loc}} c_k^\dagger | n \rangle. \tag{56}$$

The problem is to compute the local Green's function for the electron

$$\mathbf{G}_{loc}^{m,n}(z) = \langle m | c \frac{1}{z - H_{loc}} c^\dagger | n \rangle. \tag{57}$$

Let us first compute the local Green's function for $g_0 = 0$. We define diagonal Green's functions for $v_k = 0$ and then use the Dyson's equation. The diagonal Green's functions are

$$G_{o,o}^{(0)}(z) = \frac{1}{z - \varepsilon_0 - \omega_0 b^\dagger b} \tag{58}$$

and

$$G_{k,k}^{(0)}(z) = \frac{1}{z - \varepsilon_k - \omega_0 b^\dagger b}. \tag{59}$$

The Dyson's equations are

$$\begin{cases} G_{o,o}(z) = G_{o,o}^{(0)}(z) + \sum_k v_k G_{o,o}^{(0)}(z) G_{k,o}(z) \\[2ex] G_{k,o}(z) = v_k G_{k,k}^{(0)}(z) G_{o,o}(z) \end{cases} \tag{60}$$

with the solution

$$\mathbf{G}_{loc}^{(0)}(z) = G_{o,o}(z) = \frac{1}{z - \varepsilon_0 - \omega_0 b^\dagger b - W(z - \omega_0 b^\dagger b)} \tag{61}$$

and

$$W(z) = \sum_k \frac{v_k^2}{z - \varepsilon_k}. \tag{62}$$

For $g_0 \neq 0$, we consider the local Hamiltonian for the small polaron $\tilde{H}_{loc} = e^{-S} H_{loc} e^{S}$. The diagonal part is given by

$$\tilde{H}_{loc}^{(0)} = \tilde{\varepsilon}_0 c^\dagger c + \omega_0 b^\dagger b + \sum_k \varepsilon_k c_k^\dagger c_k \tag{63}$$

so we define the free Green's functions

$$\tilde{G}_{o,o}^{(0)}(z) = \frac{1}{z - \tilde{\varepsilon}_0 - \omega_0 b^\dagger b} \tag{64}$$

and

$$G_{k,k}^{(0)}(z) = \frac{1}{z - \varepsilon_k - \omega_0 b^\dagger b}. \tag{65}$$

The interaction part is given by

$$\tilde{H}_{loc}^{(1)}(z) = \sum_k v_k [U c_k^\dagger c + U^\dagger c^\dagger c_k]. \tag{66}$$

One obtains the Dyson's equations

$$\begin{cases} \tilde{G}_{o,o}(z) = \tilde{G}^{(0)}_{o,o}(z) + \sum_k v_k \tilde{G}^{(0)}_{o,o}(z) U^\dagger G_{k,o}(z) \\[2em] G_{k,o}(z) = v_k G^{(0)}_{k,k}(z) U \tilde{G}_{o,o}(z) \end{cases} \tag{67}$$

and the solution

$$\tilde{\mathbf{G}}_{loc}(z) = \tilde{G}_{o,o}(z) = \frac{1}{z - \tilde{\varepsilon}_0 - \omega_0 b^\dagger b - \tilde{\mathbf{W}}(z)} \tag{68}$$

with $\tilde{\mathbf{W}}(z) = U^\dagger W(z - \omega_0 b^\dagger b) U$. Finally, we use $\tilde{\mathbf{G}}_{loc}(z) = U^\dagger \mathbf{G}_{loc}(z) U$ and get the solution of the problem

$$\mathbf{G}_{loc}(z) = \frac{1}{z - \varepsilon_0 - \omega_0 b^\dagger b + g\omega_0 X - W(z - \omega_0 b^\dagger b)}. \tag{69}$$

We notice that $\mathbf{G}_{loc}(z) = \mathbf{G}_{at}(z - W(z - \omega_0 b^\dagger b))$. Once again, we use the Dyson's equation to compute $\mathbf{G}^{0,0}_{loc}(z)$ with

$$\mathbf{G}^{(0)}_{loc}(z) = \frac{1}{z - \varepsilon_0 - \omega_0 b^\dagger b - W(z - \omega_0 b^\dagger b)} \tag{70}$$

and

$$\mathbf{G}_{loc}(z) = \mathbf{G}^{(0)}_{loc}(z) - g_0 \omega_0 \mathbf{G}^{(0)}_{loc}(z) X \mathbf{G}_{loc}(z). \tag{71}$$

We obtain the solution

$$\mathbf{G}^{0,0}_{loc}(z) = \frac{1}{z - \varepsilon_0 - W(z) - g_0^2 \omega_0^2 \Lambda_1(z; W(z))} \tag{72}$$

with

$$\Lambda_n(z; W(z)) = \tag{73}$$
$$\frac{1}{z - \varepsilon_0 - n\omega_0 - W(z - n\omega_0) - (n+1)g_0^2 \omega_0^2 \Lambda_{n+1}(z; W(z))}.$$

Writing the solution as

$$\mathbf{G}^{0,0}_{loc}(z) = \frac{1}{z - \omega_0 - W(z) - \Sigma(z)} \tag{74}$$

we identify the self-energy.

We have thus obtained the desired local interaction self-energy $\Sigma_i(z)$ of a given site as a function only of its local Weiss's self-energy, i.e. $\Sigma_i(z) = g_0^2\omega_0^2\Lambda_1(z; W_i(z))$. Now, since we can obtain the local Weiss's self-energy as a function of the local interaction self-energies of all the other sites, we have solved the problem.

5   General Framework

Consider the Green's function of the small polaron

$$\tilde{G}_{i,j}^{\mathbf{m},\mathbf{n}}(z) = \langle\mathbf{m}|U_i^\dagger c_i \frac{1}{z-H} c_j^\dagger U_j|\mathbf{n}\rangle \tag{75}$$

with $U_j = e^{g_0(b_j^\dagger - b_j)}$. Writing the Hamiltonian $H = H^{(0)} + V$, where V is the hopping part, one obtains

$$\langle\mathbf{m}|U_i^\dagger c_i \frac{1}{z-H^{(0)}} c_j^\dagger U_j|\mathbf{n}\rangle = \frac{\delta_{i,j}\delta_{\mathbf{m},\mathbf{n}}}{z - \langle\mathbf{m}|U_j^\dagger H^{(0)} U_j|\mathbf{m}\rangle}. \tag{76}$$

We thus define the free local Green's function as

$$\tilde{G}_{i,i}^{(0)}(z) = \frac{1}{z - \tilde{\varepsilon}_0 - \omega_0 \sum_j b_j^\dagger b_j} \tag{77}$$

and get the Dyson's equation

$$\tilde{G}_{i,i}(z) = \tilde{G}_{i,i}^{(0)}(z) - t_0 \sum_\delta \tilde{G}_{i,i}^{(0)}(z) U_i^\dagger U_{i+\delta} \tilde{G}_{i+\delta,i}(z). \tag{78}$$

Now, we use the relation $\tilde{G}_{i,j}(z) = U_i^\dagger G_{i,j}(z) U_j$ to obtain the Dyson's equation for the Green's function of the electron

$$G_{i,i}(z) = G_{i,i}^{(0)}(z) - t_0 \sum_\delta G_{i,i}^{(0)}(z) G_{i+\delta,i}(z) \tag{79}$$

with

$$G_{i,i}^{(0)}(z) = \frac{1}{z - \varepsilon_0 + g_0\omega_0 X_i - \omega_0 \sum_j b_j^\dagger b_j}. \tag{80}$$

This is the Dyson's equation for a free electron on the lattice.

For a two-site problem, the local Green's functions are given by

$$G_{1,1}(z) = \frac{1}{z - \varepsilon_0 - \omega_0(b_1^\dagger b_1 + b_2^\dagger b_2) + g_0\omega_0 X_1 - t_0^2 G_{2,2}^{(0)}(z)} \tag{81}$$

and

$$\tilde{G}_{1,1}(z) = \frac{1}{z - \tilde{\varepsilon}_0 - \omega_0(b_1^\dagger b_1 + b_2^\dagger b_2) - t_0^2 U_1^\dagger G_{2,2}^{(0)}(z) U_1} \tag{82}$$

with

$$G_{2,2}^{(0)}(z) = \frac{1}{z - \varepsilon_0 - \omega_0(b_1^\dagger b_1 + b_2^\dagger b_2) + g_0\omega_0 X_2}. \tag{83}$$

We now consider a finite chain of length M. The interaction part reads

$$V = -t_0 \sum_{j=1}^{M-1} (c_j^\dagger c_{j+1} + c_{j+1}^\dagger c_j). \tag{84}$$

Consider the problem to compute the Green's function for the first site of the chain, that is $G_{1,1}(z)$, and call it, for a while, the local Green's function of the problem.

In the language of the RPE one gets

$$G_{1,1}(z) = \frac{1}{K_{1,1}^{(0)}(z) - t_0^2 G_{2,2[1]}(z)} \tag{85}$$

with $K_{i,i}^{(0)}(z) = [G_{i,i}^{(0)}(z)]^{-1}$ and $G_{2,2[1]}(z)$ the Green's function with interaction between site 1 and site 2 removed. Next, we obtain

$$G_{2,2[1]}(z) = \frac{1}{K_{2,2}^{(0)}(z) - t_0^2 G_{3,3[1,2]}(z)} \tag{86}$$

and so on, until the end of the chain,

$$G_{M,M[1,...,M-1]}(z) = G_{M,M}^{(0)}(z). \tag{87}$$

We notice that $G_{2,2[1]}(z)$ is diagonal in the basis of site 1, that $G_{3,3[1,2]}(z)$ is diagonal in the basis of site 1 and 2, and so on. This means that one can compute $G_{2,2[1]}(z)$ in the basis of all the sites but site 1. If we write $G_{2,2[1]}^{[1]}(z)$

this Green's function, then one gets $G_{2,2[1]}(z) = G^{[1]}_{2,2[1]}(z - \omega_0 b_1^\dagger b_1)$. The same way, one gets $G_{3,3[1,2]}(z) = G^{[1]}_{3,3[1,2]}(z - \omega_0 b_2^\dagger b_2)$ or $G_{3,3[1,2]}(z) = G^{[1,2]}_{3,3[1,2]}(z - \omega_0 b_1^\dagger b_1 - \omega_0 b_2^\dagger b_2)$. We also notice that $G^{[1]}_{2,2[1]}(z)$ is the local Green's function for the chain with site 1 removed, that $G^{[1,2]}_{3,3[1,2]}(z)$ is the local Green's function with sites 1 and 2 removed, and so on. The Dyson's equations become

$$G_{1,1}(z) \quad = \quad G^{(0)}_{1,1}(z) + t_0^2 G_{1,1}(z) G^{[1]}_{2,2[1]}(z - \omega_0 b_1^\dagger b_1) G^{(0)}_{1,1}(z)$$

$$G^{[1]}_{2,2[1]}(z) \quad = \quad G^{[1](0)}_{2,2}(z) + t_0^2 G^{[1]}_{2,2[1]}(z) G^{[1,2]}_{3,3[1,2]}(z - \omega_0 b_2^\dagger b_2) G^{[1](0)}_{2,2}(z)$$

$$\vdots \qquad \qquad \vdots \quad \vdots$$

$$(88)$$

Finally, we consider a Bethe lattice with coordination number Z. The Dyson's equation for the central site o is

$$G_{o,o}(z) = G^{(0)}_{o,o}(z) + t_0^2 \sum_\delta G_{o,o}(z) G_{\delta,\delta[o]}(z) G^{(0)}_{o,o}(z). \qquad (89)$$

The sum δ runs over its Z nearest neighbors. Again, $G_{\delta,\delta[o]}(z)$ is a diagonal operator in the basis of site o and we write $G_{\delta,\delta[o]}(z) = G^{[o]}_{\delta,\delta[o]}(z - \omega_0 b^\dagger b)$. Next, for one of the neighbors, the Dyson's equation reads

$$G_{\delta,\delta[o]}(z) = G^{(0)}_{\delta,\delta}(z) + t_0^2 \sum_{\delta'} G_{\delta,\delta[o]}(z) G_{\delta+\delta',\delta+\delta'[o,\delta]}(z) G^{(0)}_{\delta,\delta}(z). \qquad (90)$$

The sum δ' runs overs its $Z - 1$ nearest neigbhors (not the site o). Again $G_{\delta+\delta',\delta+\delta'[o,\delta]}(z)$ is a diagonal operator in the basis of sites o and δ.

Notice that for a Bethe lattice in infinite dimensions, we can use the cavity method [5] and obtain directly the solution of the DMFT for $\mathbf{G}_{loc}(z)$. Consider a central site o, on the lattice, with Z neighbors. Starting from the Dyson's equation (89), we develops in power of t_0^2,

$$\begin{aligned}
G_{o,o}(z) \quad = \quad & G^{(0)}_{o,o}(z) \\
& + \ t_0^2 \sum_\delta G^{(0)}_{o,o}(z) G_{\delta,\delta[o]}(z) G^{(0)}_{o,o}(z) \\
& + \ t_0^4 \sum_{\delta,\delta'} G^{(0)}_{o,o}(z) G_{\delta,\delta[o]}(z) G^{(0)}_{o,o}(z) G_{\delta',\delta'[o]}(z) G^{(0)}_{o,o}(z) \\
& + \ \cdots
\end{aligned} \qquad (91)$$

and then take the expectation value on the vacuum of all the sites, but the central site. We introduce the unknown local Green's function $\mathbf{G}_{loc}(z)$, since for $Z \to \infty$, the expectation value of $G_{o,o}(z)$ and $G_{\delta,\delta[o]}(z)$ should correspond to $\mathbf{G}_{loc}^{0,0}(z)$. At order t_0^2 one obtains the contribution to the Weiss's self-energy

$$Zt_0^2 \mathbf{G}_{loc}^{0,0}(z - \omega_0 b^\dagger b). \tag{92}$$

At order t_0^4 one obtains two contributions. The first is just

$$Z(Z-1)[t_0^2 \mathbf{G}_{loc}^{0,0}(z - \omega_0 b^\dagger b)]^2, \tag{93}$$

while the other contribution is

$$Zt_0^4 \sum_q \mathbf{G}_{loc}^{0,q}(z - \omega_0 b^\dagger b) \mathbf{G}_{at}(z - q\omega_0) \mathbf{G}_{loc}^{q,0}(z - \omega_0 b^\dagger b). \tag{94}$$

For $Z \to \infty$ and the rescaling $t_0 \to t_0/\sqrt{Z}$ only a single term survives at each step in the development in power of t_0^2 and we get the Weiss's self energy $t_0^2 \mathbf{G}_{loc}^{0,0}(z - \omega_0 b^\dagger b)$. Since, $G_{o,o}(z)$ should also correspond to $\mathbf{G}_{loc}(z)$, we obtain $\mathbf{G}_{loc}^{-1}(z) = \mathbf{G}_{at}^{-1}(z) - W(z - \omega_0 b^\dagger b)$ and the self-consistent equation $W(z) = t_0^2 \mathbf{G}_{loc}^{0,0}(z)$.

6 Revisited DMFT solution

Consider the two-site problem with the following effective Hamiltonian for site 1

$$H_1 = \varepsilon_0 c_1^\dagger c_1 + \omega_0 b_1^\dagger b_1 - g_0 \omega_0 X_1 c_1^\dagger c_1 + H_W \tag{95}$$

with

$$H_W = \varepsilon_0 c_2^\dagger c_2 - t_0(c_1^\dagger c_2 + hc) + \sum_k a_k \alpha_k^\dagger \alpha_k + \sum_k b_k(c_2^\dagger \alpha_k + hc). \tag{96}$$

The parameters a_k and b_k are choosen such that

$$\Sigma_2(z) = \sum_k \frac{b_k^2}{z - a_k}. \tag{97}$$

We use the equations of evolution to compute $G_{1,1}^{0,0}(z)$. We get

$$i\hbar \partial_t G_{1,1}^{0,0}(t) = \delta(t) + \varepsilon_0 G_{1,1}^{0,0}(t) - t_0 G_{2,1}^{0,0}(t) - g_0 \omega_0 G_{1,1}^{1,0}(t) \tag{98}$$

and

$$i\hbar\partial_t G_{1,1}^{n,0}(t) = (\varepsilon_0 + n\omega_0)G_{1,1}^{n,0}(t) - t_0 G_{2,1}^{n,0}(t) \tag{99}$$

$$-g_0\omega_0\sqrt{n}\,G_{1,1}^{n-1,0}(t) - g_0\omega_0\sqrt{n+1}\,G_{1,1}^{n+1,0}(t).$$

Now, we use the fact that

$$G_{2,1}^{n,0}(z) = \frac{-t_0 G_{1,1}^{n,0}(z)}{z - \varepsilon_0 - n\omega_0 - \Sigma_2(z - n\omega_0)} \tag{100}$$

so we get the solution

$$z G_{1,1}^{0,0}(z) = 1 + (\varepsilon_0 + \Sigma_1(z))G_{1,1}^{0,0}(z) - t_0 G_{2,1}^{0,0}(z) \tag{101}$$

or

$$G_{1,1}^{0,0}(z) = \frac{1}{z - \omega_0 - \Sigma_1(z) - W_1(z)} \tag{102}$$

with $\Sigma_1(z) = g_0^2\omega_0^2\Lambda_1(z; W_1(z))$ and

$$W_1(z) = \frac{t_0^2}{z - \varepsilon_0 - \Sigma_2(z)}. \tag{103}$$

If we use the equations of evolution to compute $G_{2,2}^{0,0}(z)$, we obtain

$$G_{2,2}^{0,0}(z) = \frac{1}{z - \varepsilon_0 - \Sigma_2(z) - \dfrac{t_0^2}{z - \varepsilon_0 - \Sigma_1(z)}}. \tag{104}$$

We thus obtain the desired self-consistent result, given by the standard solution.

For a given site o, the effective Hamiltonian is

$$H_o = \varepsilon_0 c_o^\dagger c_o + \omega_0 b_o^\dagger b_o - g_0\omega_0 X_o c_o^\dagger c_o + H_W \tag{105}$$

with

$$H_W = \varepsilon_0 \sum_{i\neq o} c_i^\dagger c_i - t_0 \sum_{i,\delta} c_{i+\delta}^\dagger c_i \tag{106}$$

$$+ \sum_{i\neq o}\sum_k [a_{k,i}\alpha_{k,i}^\dagger \alpha_{k,i} + b_{k,i}(c_i^\dagger \alpha_{k,i} + hc)]$$

The parameters $a_{k,i}$ and $b_{k,i}$ correspond to

$$\Sigma_i(z) = \sum_k \frac{b_{k,i}^2}{z - a_{k,i}} \tag{107}$$

The Hamiltonian H_W is quadratic, and correspond exactly to the Weiss's Hamiltonian of the standard theory.

Our mean field Hamiltonian H_o emphasises the fact that no excitation of phonon are allowed for the other sites of the lattice, when dealing with the excitations of a given site.

7 Restricted basis

Consider a restricted basis for the phonons, where excited states are allowed only for a single site at a time. For the two-site problem, the basis states are $|n_1,0\rangle$ and $|0,n_2\rangle$. For a three-site problem, the basis states are $|n_1,0,0\rangle$, $|0,n_2,0\rangle$, and $|0,0,n_3\rangle$. For a lattice with M sites, the dimension of the phonon basis is $(MP+1)$, if we allow P excitations for a given site.

If we compute the local Green's fonction for the site i, with the equations of evolution, we obtain the simple result, which corresponds to a local approximation,

$$zG_{i,i}^{\mathbf{0,0}}(z) = 1 + (\varepsilon_0 + \Sigma_i^{(0)}(z))G_{i,i}^{\mathbf{0,0}}(z) - t_0 \sum_\delta G_{i+\delta,i}^{\mathbf{0,0}}(z) \tag{108}$$

or

$$G_{i,i}^{\mathbf{0,0}}(z) = \frac{1}{z - \varepsilon_0 - \Sigma_i^{(0)}(z) - W_i^{(1)}(z)} \tag{109}$$

with

$$\Sigma_i^{(0)}(z) = g_0^2 \omega_0^2 \Lambda_1(z; W_i^{(0)}(z)) \tag{110}$$

and $W_i^{(0)}(z)$ is the free Weiss's self-energy, that is corresponding, in the RPE, to

$$G_{i,i}^{(0)}(z) = \frac{1}{z - \varepsilon_0} \tag{111}$$

while $W_i^{(1)}(z)$ is the Weiss's self-energy corresponding, in the RPE, to

$$G_{i,i}^{(0)}(z) = \frac{1}{z - \varepsilon_0 - \Sigma_i^{(0)}(z)}. \tag{112}$$

As an exemple, consider a one dimensional periodic lattice. First, we compute $W^{(0)}(z)$ via

$$\frac{1}{M} \sum_{\mathbf{k}} \frac{1}{z - \varepsilon_0 + 2t_0 \cos(\mathbf{k})} = \frac{1}{z - \varepsilon_0 - W^{(0)}(z)}. \qquad (113)$$

Then, we obtain $\Sigma^{(0)}(z) = g_0^2 \omega_0^2 \Lambda_1(z; W^{(0)}(z))$, and then, we get

$$\frac{1}{M} \sum_{\mathbf{k}} \frac{1}{z - \varepsilon_0 + 2t_0 \cos(\mathbf{k}) - \Sigma^{(0)}(z)} = \frac{1}{z - \varepsilon_0 - \Sigma^{(0)}(z) - W^{(1)}(z)}. \qquad (114)$$

It is interesting to notice that the expression for this local Green's function, Eq (109), corresponds to some Momentum Average Approximation [9] and to the first iteration of the DMFT solution [10].

The nice thing with this restricted basis, is that we can easily compute numerically exactly all the Green's function $G_{i,j}^{\mathbf{m,n}}(z)$ or $\tilde{G}_{i,j}^{\mathbf{m,n}}(z)$ for small systems. If we perform exact diagonalizations with P excitations, then we can compare with the analytical results by computing $\Sigma_i(z)$ also with P excitations, i.e. by limiting the continued fraction to $\Lambda_P(z; W_i(z))$. Since excitations of phonon are allowed only for a given site at a time, we expect the same structure for these Green's functions and those of the DMFT, so we can check the decoupling scheme of the Green's functions in terms of the local Green's functions.

First we can check that the local Green's functions of the problem

$$G_{i,i}^{m,n}(z) = \langle 0, \ldots m, \ldots, 0 | c_i \frac{1}{z - H} c_i^\dagger | 0, \ldots, n, \ldots, 0 \rangle \qquad (115)$$

correspond to

$$\mathbf{G}_{i,i}(z) = \frac{1}{z - \varepsilon_0 - \omega_0 b^\dagger b + g_0 \omega_0 X - \mathbf{W}_{i,i}(z)} \qquad (116)$$

where $\mathbf{W}_{i,i}(z)$ is the diagonal operator given by

$$\begin{cases} \mathbf{W}_{i,i}^{0,0}(z) = W_i^{(1)}(z) \\ \mathbf{W}_{i,i}^{n,n}(z) = W_i^{(0)}(z - n\omega_0) \ \text{ for } n > 0 \end{cases} \qquad (117)$$

We guess the form of $\mathbf{W}_{i,i}(z)$ from the fact that, when computing $\mathbf{G}_{i,i}^{0,0}(z)$, we get $W_i(z) = \mathbf{W}_{i,i}^{0,0}(z)$, while $\Sigma_i(z)$ involves only $\mathbf{W}_{i,i}^{n,n}(z)$ for $n > 0$.

Let us consider the open chain of lenght $M = 3$. For site 1 and 3 we
have

$$W_1^{(0)}(z) = \cfrac{t_0^2}{z - \varepsilon_0 - \cfrac{t_0^2}{z - \varepsilon_0}} \tag{118}$$

and for site 2 we have

$$W_2^{(0)}(z) = \frac{2t_0^2}{z - \varepsilon_0}. \tag{119}$$

From these free Weiss's self-energies, we compute

$$\Sigma_1^{(0)}(z) = g_0^2 \omega_0^2 \Lambda_1(z; W_1^{(0)}(z)) \tag{120}$$

and

$$\Sigma_2^{(0)}(z) = g_0^2 \omega_0^2 \Lambda_1(z; W_2^{(0)}(z)). \tag{121}$$

Then we get

$$W_1^{(1)}(z) = \cfrac{t_0^2}{z - \varepsilon_0 - \Sigma_2^{(0)}(z) - \cfrac{t_0^2}{z - \varepsilon_0 - \Sigma_1^{(0)}(z)}} \tag{122}$$

and

$$W_2^{(1)}(z) = \frac{2t_0^2}{z - \varepsilon_0 - \Sigma_1^{(0)}(z)}. \tag{123}$$

Then we compute $G_{1,1}^{\mathbf{0,0}}(z)$ and $G_{2,2}^{\mathbf{0,0}}(z)$ with our local approximation

$$\mathbf{G}_{1,1}^{\mathbf{0,0}}(z) = \frac{1}{z - \varepsilon_0 - \Sigma_1^{(0)}(z) - W_1^{(1)}(z)} \tag{124}$$

and

$$\mathbf{G}_{2,2}^{\mathbf{0,0}}(z) = \frac{1}{z - \varepsilon_0 - \Sigma_2^{(0)}(z) - W_2^{(1)}(z)}. \tag{125}$$

We obtain a perfect agreement with the exact result.

Next we compute $G_{1,2}^{\mathbf{0,0}}(z)$. We start from the exact expression

$$G_{1,2}(z) = -t_0 G_{1,1}(z) G_{2,2[1]}(z) \tag{126}$$

and make the decoupling

$$G_{1,2}^{\mathbf{0,0}}(z) = -t_0\mathbf{G}_{1,1}^{0,0}(z)\mathbf{G}_{2,2[1]}^{0,0}(z). \tag{127}$$

We guess that the local Green's function $\mathbf{G}_{2,2[1]}(z)$ is given by

$$\mathbf{G}_{2,2[1]}(z) = \frac{1}{z - \varepsilon_0 - \omega_0 b^\dagger b + g_0\omega_0 X - \mathbf{W}_{2,2[1]}(z)} \tag{128}$$

with the diagonal operator

$$\begin{cases} \mathbf{W}_{2,2[1]}^{0,0}(z) = W_{2[1]}^{(1)}(z) \\ \mathbf{W}_{2,2[1]}^{n,n}(z) = W_2^{(0)}(z - n\omega_0) \text{ for } n > 0 \end{cases} \tag{129}$$

and the Weiss's function

$$W_{2[1]}^{(1)}(z) = \frac{t_0^2}{z - \varepsilon_0 - \Sigma_3^{(0)}(z)}. \tag{130}$$

We thus obtain

$$\mathbf{G}_{2,2[1]}^{0,0}(z) = \frac{1}{z - \varepsilon_0 - \Sigma_2^{(0)}(z) - W_{2[1]}^{(1)}(z)}. \tag{131}$$

We obtain a perfect agreement with the exact result.

Next we compute $G_{1,3}^{\mathbf{0,0}}(z)$. We start from the exact expression

$$G_{1,3}(z) = -t_0 G_{1,2}(z)G_{3,3[1,2]}(z) \tag{132}$$

and make the decoupling

$$G_{1,3}^{\mathbf{0,0}}(z) = +t_0^2\mathbf{G}_{1,1}^{0,0}(z)\mathbf{G}_{2,2[1]}^{0,0}(z)\mathbf{G}_{3,3[1,2]}^{0,0}(z). \tag{133}$$

We Guess that $\mathbf{G}_{3,3[1,2]}(z)$ is given by

$$\mathbf{G}_{3,3[1,2]}(z) = \frac{1}{z - \varepsilon_0 - \omega_0 b^\dagger b + g_0\omega_0 X - \mathbf{W}_{3,3[1,2]}(z)} \tag{134}$$

with the diagonal operator

$$\begin{cases} \mathbf{W}_{3,3[1,2]}^{0,0}(z) = 0 \\ \mathbf{W}_{3,3[1,2]}^{n,n}(z) = W_3^{(0)}(z - n\omega_0) \text{ for } n > 0 \end{cases}. \tag{135}$$

We get

$$\mathbf{G}^{0,0}_{33[1,2]}(z) = \frac{1}{z - \varepsilon_0 - \Sigma^{(0)}_3(z)}. \tag{136}$$

We obtain a perfect agreement with the exact result.

For $\tilde{G}^{0,0}_{i,i}(z)$, we compute

$$\tilde{G}^{0,0}_{1,1}(z) = \sum_m \sum_n U^{m,0} \mathbf{G}^{m,n}_{1,1}(z) U^{n,0} \tag{137}$$

and

$$\tilde{G}^{0,0}_{2,2}(z) = \sum_m \sum_n U^{m,0} \mathbf{G}^{m,n}_{2,2}(z) U^{n,0}. \tag{138}$$

We obtain a perfect agreement with the exact results.

For $\tilde{G}^{0,0}_{i,j}(z)$, we compute

$$\tilde{G}^{0,0}_{1,2}(z) = -t_0 \sum_m \sum_n U^{m,0} \mathbf{G}^{m,0}_{1,1}(z) \mathbf{G}^{0,n}_{2,2[1]}(z) U^{n,0} \tag{139}$$

and

$$\tilde{G}^{0,0}_{1,3}(z) = t_0^2 \sum_m \sum_n U^{m,0} \mathbf{G}^{m,0}_{1,1}(z) \mathbf{G}^{0,0}_{2,2[1]}(z) \mathbf{G}^{0,n}_{3,3[1,2]}(z) U^{n,0}. \tag{140}$$

We obtain a perfect agreement with the exact results.

8 Decoupling scheme

We apply the decoupling scheme obtained within the restricted basis to compute $G^{0,0}_{i,j}(z)$ and $\tilde{G}^{0,0}_{i,j}(z)$, within the DMFT, that is, instead to deal with $W^{(0)}_i(z)$ and $W^{(1)}_i(z)$, we use the self-consistent solution $W_i(z)$ given by the DMFT.

Consider a lattice with M sites and solve the DMFT equations to get $\Sigma_i(z)$ and $W_i(z)$. Then construct all the local Green's functions

$$\mathbf{G}^{(0)}_{i,i}(z) = \frac{1}{z - \varepsilon_0 - \omega_0 b^\dagger b + g_0 \omega_0 X - \mathbf{W}_{i,i}(z)} \tag{141}$$

with the diagonal operator

$$\left\{ \begin{array}{l} \mathbf{W}^{0,0}_{i,i}(z) = 0 \\ \mathbf{W}^{n,n}_{i,i}(z) = W_i(z - n\omega_0) \ \text{ for } n > 0 \end{array} \right. \tag{142}$$

Then use the RPE to express any Green's function in term of these local Green's functions and take the expectation value individually.

For the electron, the RPE is just the equation of free particle on the lattice, so we just need

$$G_{i,i}^{(0)}(z) = [\mathbf{G}_{i,i}^{(0)}(z)]^{0,0} = \frac{1}{z - \varepsilon_0 - \Sigma_i(z)}. \tag{143}$$

For an infinite system, $\Sigma_i(z) = \Sigma(z)$, and for a periodic lattice, we obtain

$$G(\mathbf{k}, z) = \frac{1}{z - \varepsilon_0 - \xi_\mathbf{k} - \Sigma(z)}. \tag{144}$$

For the small polaron, the RPE involve the operators U_i, so we need to compute $[U^\dagger \mathbf{G}_{i,i}^{(0)}(z)]^{0,0} = [\mathbf{G}_{i,i}^{(0)}(z)U]^{0,0}$ or $[U^\dagger \mathbf{G}_{i,i}^{(0)}(z)U]^{0,0}$.

Let us compute $G(\mathbf{k}, z)$ and $\tilde{G}(\mathbf{k}, z)$ for a one-dimensional lattice with M sites. In this case, we have $\Sigma_i(z) = \Sigma(z)$ and $W_i(z) = W(z)$. We start with

$$W_{M[1,\dots,M-1]}(z) = 0 \tag{145}$$

then compute

$$W_{i[1,\dots,i-1]}(z) = \frac{t_0^2}{z - \varepsilon_0 - \Sigma(z) - W_{i+1[1,\dots,i]}(z)} \tag{146}$$

until $W_{2[1]}(z)$, and then

$$W_1(z) = W(z) = \frac{2t_0^2}{z - \varepsilon_0 - \Sigma(z) - W_{2[1]}(z)}. \tag{147}$$

We solve the self-consistent equations for $\Sigma(z) = g_0^2 \omega_0^2 \Lambda_1(z; W(z))$ and $W(z)$. Notice that this way to compute $W(z)$, instead of Eq (114), gives a slightly different result that vanishes for M large enough.

In order to compute $G(\mathbf{k}, z)$, we just have to consider

$$\mathbf{G}_{i,i[1,\dots,i-1]}^{0,0}(z) = \frac{1}{z - \varepsilon_0 - \Sigma(z) - W_{i[1,\dots,i-1]}(z)}. \tag{148}$$

We start to compute the local Green's function $G_{1,1}^{0,0}(z)$ that is $\mathbf{G}_{1,1}^{0,0}(z)$, and then compute

$$G_{1,j+1}^{0,0}(z) = -t_0 G_{1,j}^{0,0}(z) \mathbf{G}_{j+1,j+1[1,\dots,j]}^{0,0}(z) \tag{149}$$

until $G_{1,M}^{\mathbf{0,0}}(z)$. Then we use the Fourier transform. From a numerical point of view, it is safer to choose M odd and to compute

$$G(k,z) = G_{o,o}^{\mathbf{0,0}}(z) + \sum_{n=1}^{(M+1)/2} 2\cos(2\pi kn/M)\, G_{o,n}^{\mathbf{0,0}}(z) \qquad (150)$$

In this case, we can compare the numerical result with the standard result of the DMFT, i.e. Eq (144).

Next, for $\tilde{G}(\mathbf{k}, z)$, we have to compute the full local operators

$$\mathbf{G}_{i,i[1,\ldots,i-1]}(z) = \frac{1}{z - \varepsilon_0 - \omega_0 b^\dagger b + g_0\omega_0 X - \mathbf{W}_{i,i[1,\ldots,i-1]}(z)} \qquad (151)$$

with

$$\begin{cases} \mathbf{W}_{i,i[1,\ldots,i-1]}^{0,0}(z) = W_{i[1,\ldots,i-1]}(z) \\[2ex] \mathbf{W}_{i,i[1,\ldots,i-1]}^{n,n}(z) = W(z - n\omega_0) \text{ for } n > 0 \end{cases} \qquad (152)$$

Then we compute

$$\tilde{G}_{1,1}^{\mathbf{0,0}}(z) = \sum_{m,n} U^{m,0}U^{n,0}\mathbf{G}_{1,1}^{m,n}(z), \qquad (153)$$

$$\tilde{G}_{1,2}^{\mathbf{0,0}}(z) = -t_0 \sum_{m,n} U^{m,0}U^{n,0}\mathbf{G}_{1,1}^{m,0}(z)\mathbf{G}_{2,2[1]}^{0,n}(z), \qquad (154)$$

$$\tilde{G}_{1,3}^{\mathbf{0,0}}(z) = t_0^2 \sum_{m,n} U^{m,0}U^{n,0}\mathbf{G}_{1,1}^{m,0}(z)\mathbf{G}_{2,2[1]}^{0,0}(z)\mathbf{G}_{3,3[1,2]}^{0,m}(z), \qquad (155)$$

$$\tilde{G}_{1,4}^{\mathbf{0,0}}(z) = -t_0^3 \sum_{m,n} U^{m,0}U^{n,0}\mathbf{G}_{1,1}^{m,0}(z)\mathbf{G}_{2,2[1]}^{0,0}(z)\mathbf{G}_{3,3[1,2]}^{0,0}(z)\mathbf{G}_{4,4[1,2,3]}^{0,m}(z),$$
$$\qquad (156)$$

and so on, until $\tilde{G}_{1,M}^{\mathbf{0,0}}(z)$. Then we get

$$\tilde{G}(k,z) = \tilde{G}_{o,o}^{\mathbf{0,0}}(z) + \sum_{n=1}^{(M+1)/2} 2\cos(2\pi kn/M)\, \tilde{G}_{o,n}^{\mathbf{0,0}}(z). \qquad (157)$$

In this case, we can check that the sum over $\mathbf{k}$ corresponds to the local Green's function

$$\tilde{G}_{1,1}^{\mathbf{0,0}}(z) = \frac{1}{M} \sum_{k} \tilde{G}(k,z). \qquad (158)$$

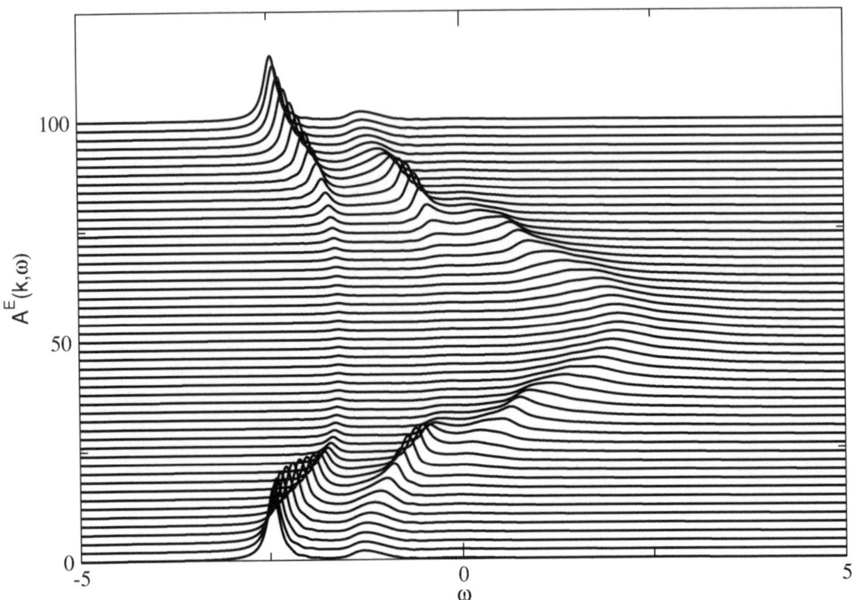

Figure 2: Spectral functions for the electron

In Fig 2 and Fig 3 we show the result for the spectral functions of the electron

$$A(k,\omega) = -2\,\mathcal{I}m G(k, z = \omega + i\eta) \tag{159}$$

and for the spectral functions of the small polaron

$$\tilde{A}(k,\omega) = -2\,\mathcal{I}m\tilde{G}(k, z = \omega + i\eta) \tag{160}$$

The parameters are $t_0 = \omega_0 = g_0 = 1$, $\varepsilon_0 = 0$, $M = 51$, and $\eta = 0.1$.

For these values of the parameters, we obtain that $A(k = 0,\omega)$ and $\tilde{A}(k = 0,\omega)$ have exactly the same main pic, but we obtain a rather different physics for the electron and the small polaron. The small polaron of the atomic limit remains a good quasiparticle.

9 Dressing an electron

We would like to emphasis that they are many ways to define a small polaron operator, i.e. to dress localy an electron with phonons [4].

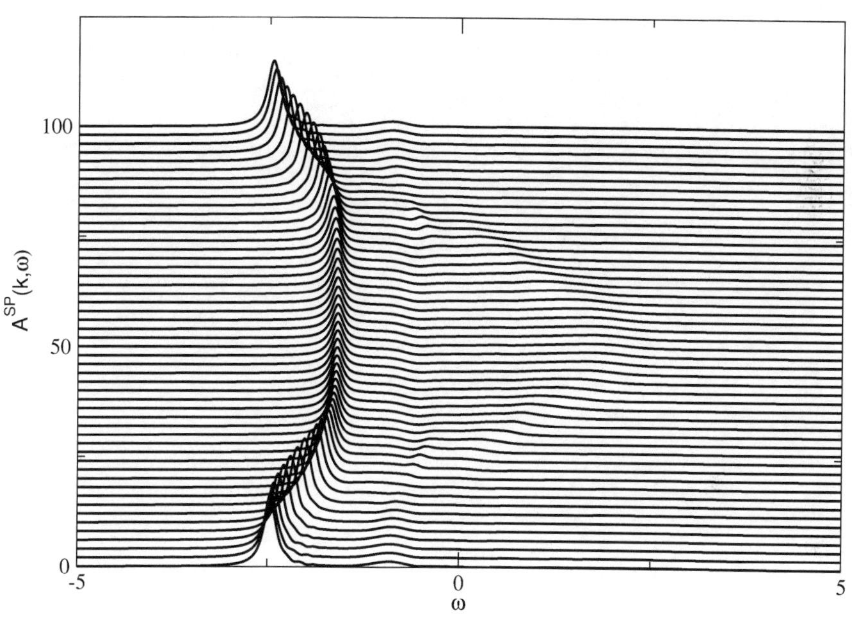

Figure 3: Spectral functions for the small polaron

Consider the two-site problem and compute the Green's functions for the electron

$$G_{1,1}^{m,n}(z) = \langle m,0|c_1 \frac{1}{z-H} c_1^\dagger |n,0\rangle \tag{161}$$

and

$$G_{1,2}^{m,n}(z) = \langle m,0|c_1 \frac{1}{z-H} c_2^\dagger |0,n\rangle. \tag{162}$$

If we use the operator U to compute the Green's functions for the small polaron $\tilde{G}_{1,1}^{00,00}(z)$ and $\tilde{G}_{1,2}^{00,00}(z)$, with

$$\tilde{G}_{i,j}^{00,00}(z) = \sum_{m,n} U^{m,0} G_{i,j}^{m,n}(z) U^{n,0} \tag{163}$$

we obtain, for the first poles, that correspond to the fundamental energy and the first excited state,

$$\begin{array}{cc} E_p & \mathcal{Z}_p \\ -1.68848 & 0.43015 \\ -1.14106 & \pm 0.37575 \end{array} \tag{164}$$

So $\tilde{G}(k=0,z)$ has a main pic with $\mathcal{Z} \simeq 0.86$ while $\tilde{G}(k=\pi,z)$ has a main pic with $\mathcal{Z} \simeq 0.75$.

Next, following Ref [4], we can also define a small polaron using the reduced density matrix for a single site. We have, for the eigenstates,

$$|p) = \sum_{j,m_1,m_2} \Psi_{j,m_1,m_2}(p) \, c_j^\dagger |m_1,m_2\rangle \tag{165}$$

so, targetting only on one of the eigenstates, we obtain the reduced density matrix

$$\rho^{m,n}(p) = \sum_{m_2} \Psi^*_{1,m,m_2}(p) \Psi_{1,n,m_2}(p) \tag{166}$$

Then, we diagonalize $\rho(p)$ and get $U(p)$. The first eigenstate with $E_0 \simeq -1.688$ corresponds to the symmetry $\mathbf{k}=0$ and the next with $E_1 \simeq -1.141$, corresponds to the symmetry $\mathbf{k}=\pi$. For $U(0)$, we get the result

$$\begin{array}{cc} E_p & \mathcal{Z}_p \\ -1.68848 & 0.46065 \\ -1.14106 & \pm 0.28247 \end{array} \tag{167}$$

So $\tilde{G}(k = 0, z)$ has a main pic with $\mathcal{Z} \simeq 0.92$ while $\tilde{G}(k = \pi, z)$ has a main pic with $\mathcal{Z} \simeq 0.57$. For $U(\pi)$, we get the result

$$
\begin{array}{cc}
E_p & \mathcal{Z}_p \\
-1.68848 & 0.35216 \\
-1.14106 & \pm 0.41841
\end{array}
\tag{168}
$$

So, this time, $\tilde{G}(k = 0, z)$ has a main pic with $\mathcal{Z} \simeq 0.70$ while $\tilde{G}(k = \pi, z)$ has a main pic with $\mathcal{Z} \simeq 0.84$.

The formalism to compute the non local Green's functions of the small polaron within the Dynamical Mean Field Theory, presented in the last section, can be used with any local transformation.

10   Conclusion

In this paper we revisited the Dynamical Mean Field Theory of the Holstein Polaron Problem in order to compute the all dynamical correlation functions for the electron and the small polaron. We introduced a restricted basis for the phonons in order to check numerically the right decoupling scheme of the Green's functions within a local approximation.

References

[1] T. Holstein, Ann. Phys. (N.Y.) **8**, 325 (1959).

[2] H. Fehske and S. A. Trugman, in *Polarons in Advanced Materials*, edited by A. S. Alexandrov, Springer Series in Material Sciences 103 pp. 393-461, Springer Verlag, Dordrecht (2007).

[3] J. M. Robin, Phys. Rev. B **58**, 14335 (1998).

[4] C. Zhang, E. Jeckelmann, and S. R. White, Phys. Rev. B **60**, 14092 (1999).

[5] A. Georges, G. Kotliar, W. Krauth, and M. J. Rozenberg, Rev. Mod. Phys. **68**, 13-125 (1996).

[6] S. Ciuchi, F. de Pasquale, S. Fratini, and D. Feinberg, Phys. Rev. B **56**, 4494 (1997).

[7] E. N. Economou, *Green's Function in Quantum Physics*, Springer-Verlage, Berlin.

[8] H. Haken, *Quantum field theory of solids*, North Holland Publishing, Amsterdam, 1976.

[9] M. Berciu, Phys. Rev. Lett. **97**, 036402 (2006).

[10] O. S. Barisic, Phys. Rev. B **76**, 193106 (2007).

Numerical Stuff

Local self-energy

We make the program `berciu.cc` to compute the local self-energy for a system with M sites. First, we compute the free local Green's function

$$G_0(z) = \frac{1}{M} \sum_{k=0}^{M-1} \frac{t_0^2}{z - \varepsilon_0 - \xi_k} \tag{169}$$

with

$$\xi_k = -2t_0 \cos\left(\frac{2\pi}{M}k\right) \tag{170}$$

and then extract the free Weiss's function using the formula

$$G_0(z) = \frac{1}{z - \varepsilon_0 - W_0(z)} \tag{171}$$

Then we compute the self-energy

$$\Sigma_0(z) = g_0^2 \omega_0^2 \Lambda_1(z) \tag{172}$$

with

$$\Lambda_n(z) = \frac{1}{z - \varepsilon_0 - n\omega_0 - W_0(z - n\omega_0) - (n+1)g_0^2 \omega_0^2 \Lambda_{n+1}(z)} \tag{173}$$

Then we can compute

$$G(k, z) = \frac{1}{z - \varepsilon_0 - \xi_k - \Sigma_0(z)} \tag{174}$$

or

$$G(z) = \frac{1}{M} \sum_k G(k, z) \tag{175}$$

and extract $W_1(z)$ such that

$$G(z) = \frac{1}{z - \varepsilon_0 - \Sigma_0(z) - W_1(z)} \tag{176}$$

We compare the results $G(k = 0, z)$ with those in the paper of Berciu to check our program.

DMFT

We start with $W(z) = W_0(z)$, then compute $\Sigma(z)$ for $W(z)$ given, then compute $G(z)$ and extract $W(z)$ for $\Sigma(z)$ given and so on.

We write the program `dmft.cc` to do the job. We obtain, for our parameters, the same result for the lowest first pic.

We notice that the Berciu approximation is not a self-consistent one. If we write the equations of motion, we obtain

$$z G_{ii}^{0,0}(z) = 1 + \varepsilon_0 G_{ii}^{0,0}(z) + \Sigma_0(z) G_{ii}^{0,0}(z) - t_0 G_{i+\delta,i}^{0,0}(z) \tag{177}$$

and

$$z G_{i+\delta,i}^{0,0}(z) = \varepsilon_0 G_{i+\delta,i}^{0,0}(z) + \Sigma_0(z) G_{i+\delta,i}^{0,0}(z) - t_0 G_{i+\delta+\delta',i}^{0,0}(z) \tag{178}$$

Thus our local Green's function corresponds to

$$G_{ii}^{0,0}(z) = \frac{1}{z - \varepsilon_0 - \Sigma_0(z) - W_1(z)} \tag{179}$$

The local Green's function of the DMFT corresponds to

$$G_{ii}^{0,0}(z) = \frac{1}{z - \varepsilon_0 - \Sigma_\infty(z) - W_\infty(z)} \tag{180}$$

However, in next section, we see how to construct the local Green's function.

Exact diagonalizations

We make a program `exact.cc` to compute the local Green's function. We allow P excitations of phonon for each site. For $P = 2$ and $M = 4$ the basis for the phonons is

i	n_1	n_2	n_3	n_4
0	0	0	0	0
1	1	0	0	0
2	2	0	0	0
3	0	1	0	0
4	0	2	0	0
5	0	0	1	0
6	0	0	2	0
7	0	0	0	1
8	0	0	0	2

$$(181)$$

The dimension of the basis is $(MP + 1)$. We then construct

$$H(n) = \varepsilon_0 + \omega_0 \sum_j b_j^\dagger b_j - g_0 \omega_0 X_n \tag{182}$$

We compute $G_{11}(z)$. We obtain a perfect agreement with the local Green's function in the Berciu's approximation.

However, we write the program `berciu2.cc` to compute $W_0(z)$ with the RPE method and find out that only the low lying part of the spectrum agree. For $M > 50$, there is a very good agreement.

We compute the local Green's function for the small polaron $\tilde{G}_{11}^{0,0}(z)$.

$$\tilde{G}_{11}(z) = \sum_{n=0}^{P} \sum_{m=0}^{P} \sum_{\lambda} \frac{U^{n,0} V^{n,\lambda} U^{m,0} V^{m,\lambda}}{z - E_\lambda} \tag{183}$$

with

$$U^{n,0} = e^{-g_0^2/2} \frac{g_0^n}{\sqrt{n!}} \tag{184}$$

In order to compare this result with the local theory, we write the program `local.cc`. First, we compute the local Green's function

$$\mathbf{G}_{loc}(z) = \frac{1}{z - \varepsilon_0 - \omega_0 b^\dagger b + g_0 \omega_0 X - W_{loc}(z - \omega_0 b^\dagger b)} \tag{185}$$

The function $W_{loc}(z)$ should be $W_1(z)$ in order to get the Weiss's self-energy and $W_0(z)$ in order to give $\Sigma_0(z)$, so this is just

$$[W_{loc}(z - \omega_0 b^\dagger b)]^{0,0} = W_1(z) \tag{186}$$

and, for $n > 0$,

$$[W_{loc}(z - \omega_0 b^\dagger b)]^{n,n} = W_0(z - n\omega_0) \tag{187}$$

We check that we obtain the right result for $\mathbf{G}_{loc}^{0,0}(z)$ and then we compute

$$\tilde{\mathbf{G}}_{loc}^{0,0}(z) = \sum_{m=0}^{P} \sum_{n=0}^{P} U^{m,0} U^{n,0} \mathbf{G}_{loc}^{m,n}(z) \tag{188}$$

We obtain a perfect agreement with the exact diagonalization result.

Two-site problem

We write the program `exact2.cc` to compute

$$G_{12}^{m0,0n}(z) \tag{189}$$

From the equations of evolution, we get

$$G_{12}^{00,00}(z) = \frac{-t_0}{z - \varepsilon_0 - \Sigma_0(z)} \frac{1}{z - \varepsilon_0 - \Sigma_0(z) - W_1(z)} \tag{190}$$

with

$$W_0(z) = \frac{t_0^2}{z - \varepsilon_0} \tag{191}$$

$$\Sigma_0(z) = g_0^2 \omega_0^2 \Lambda_1(z; W_0(z)) \tag{192}$$

and

$$W_1(z) = \frac{t_0^2}{z - \varepsilon_0 - \Sigma_0(z)} \tag{193}$$

We thus write the program `twoG.cc` to compare the result. We obtain a perfect agreement.

Next, we use our programm `exact2.cc` to compute

$$\tilde{G}_{12}^{00,00}(z) = \sum_{m=0}^{P} \sum_{n=0}^{P} U^{m,0} U^{n,0} G_{12}^{m0,0n}(z) \tag{194}$$

We check that we obtain the right answer for $\tilde{G}_{11}^{00,00}(z)$ and $\tilde{G}_{22}^{00,00}(z)$.

Then we write the program `twoP.cc`. The idea is to start from the exact solution

$$G_{12}(z) = -t_0 G_{11}(z) G_{22[1]}(z) \tag{195}$$

Within the local approximation, we expect the decoupling scheme

$$G_{12}^{m0,0n}(z) = -t_0 \mathbf{G}_{11}^{m,0}(z) \mathbf{G}_{22[1]}^{0,n}(z) \tag{196}$$

with

$$\mathbf{G}_{11}(z) = \frac{1}{z - \varepsilon_0 - \omega_0 b_1^\dagger b_1 + g_0 \omega_0 X_1 - W_{11}(z)} \tag{197}$$

and

$$\mathbf{G}_{22[1]}(z) = \frac{1}{z - \varepsilon_0 - \omega_0 b_2^\dagger b_2 + g_0 \omega_0 X_2 - W_{22[1]}(z)} \tag{198}$$

with the diagonal operators

$$\begin{aligned} W_{11}^{0,0}(z) &= W_1(z) \\ W_{11}^{n,n}(z) &= W_0(z - n\omega_0) \end{aligned} \tag{199}$$

and

$$\begin{aligned} W_{22[1]}^{0,0}(z) &= 0 \\ W_{22[1]}^{n,n}(z) &= W_0(z - n\omega_0) \end{aligned} \tag{200}$$

We thus compute $\mathbf{G}_{11}(z)$ and $\mathbf{G}_{22[1]}(z)$ and then compute

$$\tilde{G}_{12}^{00,00}(z) = -t_0 \sum_m \sum_n U^{m,0} \mathbf{G}_{11}^{m,0}(z) \mathbf{G}_{22[1]}^{0,n}(z) U^{n,0} \tag{201}$$

We obtain a perfect agreement.

We apply our idea to the DMFT for the two-site problem. We write the program `dmft2.cc` to compute $W(z)$ and $\Sigma(z)$. We obtain

$$G_{11}^{00,00}(z) = \frac{1}{z - \varepsilon_0 - \Sigma(z) - W(z)} \tag{202}$$

and

$$G_{12}^{00,00}(z) = -t_0 W(z) G_{11}(z) \tag{203}$$

We construct

$$\mathbf{G}_{ii}(z) = \frac{1}{z - \varepsilon_0 - \omega_0 b^\dagger b + g_0 \omega_0 X - W_{ii}(z)} \tag{204}$$

with

$$\begin{aligned} W_{11}^{0,0}(z) &= W(z) \\ W_{11}^{n,n}(z) &= W(z - n\omega_0) \end{aligned} \tag{205}$$

and

$$\begin{aligned} W_{22[1]}^{0,0}(z) &= 0 \\ W_{22[1]}^{n,n}(z) &= W(z - n\omega_0) \end{aligned} \tag{206}$$

We compute

$$G_{12}^{00,00}(z) = -t_0 \mathbf{G}_{11}^{0,0}(z) \mathbf{G}_{22[1]}^{0,0}(z) \tag{207}$$

We obtain a perfect agreement.

As a bonus, we compute

$$\tilde{G}_{12}^{00,00}(z) = -t_0 \sum_m \sum_n U^{m,0} \mathbf{G}_{11}^{m,0}(z) \mathbf{G}_{22[1]}^{0,n} U^{n,0} \tag{208}$$

A good check is to compute

$$\tilde{G}_{k=0}(z) = \tilde{G}_{11}(z) + \tilde{G}_{12}(z) \tag{209}$$

and

$$\tilde{G}_{k=\pi}(z) = \tilde{G}_{11}(z) - \tilde{G}_{12}(z) \tag{210}$$

We obtain two well defined spectral functions.

Three-site problem

We consider an open chain with three sites.

We write the program `exact3.cc`.

We compute $G_{11}(z)$, $G_{12}(z)$, $G_{22}(z)$, $G_{13}(z)$, for the electron; $\tilde{G}_{11}(z)$, $\tilde{G}_{12}(z)$, $\tilde{G}_{22}(z)$, $\tilde{G}_{13}(z)$, for the small polaron.

We write the program `threeG.cc` to compute $G_{ij}(z)$. For site 1 and 3 we have

$$W_1^{(0)}(z) = \frac{t_0^2}{z - \varepsilon_0 - \dfrac{t_0^2}{z - \varepsilon_0}} \tag{211}$$

For site 2 we have

$$W_2^{(0)}(z) = \frac{2t_0^2}{z - \varepsilon_0} \tag{212}$$

From these free Weiss's self-energies, we compute

$$\Sigma_1^{(0)}(z) = g_0^2\omega_0^2\Lambda(z; W_1^{(0)}(z)) \tag{213}$$

and

$$\Sigma_2^{(0)}(z) = g_0^2\omega_0^2\Lambda(z; W_2^{(0)}(z)) \tag{214}$$

Then we get

$$W_1^{(1)}(z) = \frac{t_0^2}{z - \varepsilon_0 - \Sigma_2^{(0)}(z) - \dfrac{t_0^2}{z - \varepsilon_0 - \Sigma_1^{(0)}(z)}} \tag{215}$$

and

$$W_2^{(1)}(z) = \frac{2t_0^2}{z - \varepsilon_0 - \Sigma_1^{(0)}(z)} \tag{216}$$

We compute

$$G_{11}^{0,0}(z) = \frac{1}{z - \varepsilon_0 - \Sigma_1^{(0)}(z) - W_1^{(1)}(z)} \tag{217}$$

and

$$G_{22}^{0,0}(z) = \frac{1}{z - \varepsilon_0 - \Sigma_2^{(0)}(z) - W_2^{(1)}(z)} \tag{218}$$

We obtain a perfect agreement.

Next we compute $G_{12}(z)$. We start from the exact relation

$$G_{12}(z) = -t_0 G_{11}(z)G_{22[1]}(z) \tag{219}$$

We see we have to compute

$$W_{2[1]}^{(1)}(z) = \frac{t_0^2}{z - \varepsilon_0 - \Sigma_1^{(0)}(z)} \tag{220}$$

So we obtain

$$G_{12}^{0,0}(z) = -t_0 \mathbf{G}_{11}^{0,0}(z)\mathbf{G}_{22[1]}^{0,0}(z) \tag{221}$$

with

$$\mathbf{G}^{0,0}_{22[1]}(z) = \frac{1}{z - \varepsilon_0 - \Sigma^{(0)}_2(z) - W^{(1)}_{2[1]}(z)} \tag{222}$$

We obtain a perfect agreement.

Next we compute $G^{0,0}_{13}(z)$. We start from the exact expression

$$G_{13}(z) = -t_0 G_{12}(z) G_{33[1,2]}(z) \tag{223}$$

So we have to compute

$$\mathbf{G}^{0,0}_{33[1,2]}(z) = \frac{1}{z - \varepsilon_0 - \Sigma^{(0)}_1(z)} \tag{224}$$

We obtain a perfect agreement.

We write a program `threeP.cc` to compute $\tilde{G}^{0,0}_{ij}(z)$. For $\tilde{G}^{0,0}_{11}(z)$, we build

$$\mathbf{G}_{11}(z) = \frac{1}{z - \varepsilon_0 - \omega_0 b^\dagger b + g_0 \omega_0 X - W_{11}(z)} \tag{225}$$

The diagonal Weiss's operator is

$$\begin{aligned}
W^{0,0}_{11}(z) &= W^{(1)}_1(z) \\
W^{n,n}_{11}(z) &= W^{(0)}_1(z - n\omega_0)
\end{aligned} \tag{226}$$

Then we compute

$$\tilde{G}^{0,0}_{11}(z) = \sum_m \sum_n U^{m,0} \mathbf{G}^{m,n}_{11}(z) U^{n,0} \tag{227}$$

We obtain a perfect agreement.

For $\tilde{G}^{0,0}_{22}(z)$, we build

$$\mathbf{G}_{22}(z) = \frac{1}{z - \varepsilon_0 - \omega_0 b^\dagger b + g_0 \omega_0 X - W_{22}(z)} \tag{228}$$

The diagonal Weiss's operator is

$$\begin{aligned}
W^{0,0}_{22}(z) &= W^{(1)}_2(z) \\
W^{n,n}_{22}(z) &= W^{(0)}_2(z - n\omega_0)
\end{aligned} \tag{229}$$

Then we compute

$$\tilde{G}_{22}^{\mathbf{0,0}}(z) = \sum_m \sum_n U^{m,0} \mathbf{G}_{22}^{m,n}(z) U^{n,0} \qquad (230)$$

We obtain a perfect agreement.

For $\tilde{G}_{12}^{\mathbf{0,0}}(z)$, we build

$$\mathbf{G}_{22[1]}(z) = \frac{1}{z - \varepsilon_0 - \omega_0 b^\dagger b + g_0 \omega_0 X - W_{22[1]}(z)} \qquad (231)$$

The diagonal Weiss's operator is

$$\begin{aligned} W_{22[1]}^{0,0}(z) &= W_{2[1]}^{(1)}(z) \\ W_{22[1]}^{n,n}(z) &= W_2^{(0)}(z - n\omega_0) \end{aligned} \qquad (232)$$

Then we compute

$$\tilde{G}_{12}^{\mathbf{0,0}}(z) = -t_0 \sum_m \sum_n U^{m,0} \mathbf{G}_{11}^{m,0}(z) \mathbf{G}_{22[1]}^{0,n}(z) U^{n,0} \qquad (233)$$

We obtain a perfect agreement.

For $\tilde{G}_{13}^{\mathbf{0,0}}(z)$, we build

$$\mathbf{G}_{33[1,2]}(z) = \frac{1}{z - \varepsilon_0 - \omega_0 b^\dagger b + g_0 \omega_0 X - W_{33[1,2]}(z)} \qquad (234)$$

The diagonal Weiss's operator is

$$\begin{aligned} W_{33[1,2]}^{0,0}(z) &= 0 \\ W_{33[1,2]}^{n,n}(z) &= W_3^{(0)}(z - n\omega_0) \end{aligned} \qquad (235)$$

Then we compute

$$\tilde{G}_{13}^{\mathbf{0,0}}(z) = t_0^2 \sum_m \sum_n U^{m,0} \mathbf{G}_{11}^{m,0}(z) \mathbf{G}_{22[1]}^{0,0}(z) \mathbf{G}_{33[1,2]}^{0,n}(z) U^{n,0} \qquad (236)$$

We obtain a perfect agreement.

Dispersion

We want to compute $G(\mathbf{k}, z)$ and $\tilde{G}(\mathbf{k}, z)$ for a one dimensional lattice within the DMFT.

We write the program `dmft11.cc` for a closed chain of M sites and we will save the first 11 spectral functions. First we compute $\Sigma(z)$ and $W(z)$. We consider the two possible ways to compute $W(z)$. Then we compute

$$G(k, z) = \frac{1}{z - \varepsilon_0 - \xi_k - \Sigma(z)} \tag{237}$$

Then we save the spectral functions $A(k, \omega)$.

We write the program `disp11e.cc` for a closed chain of M sites and we will save the first 11 spectral functions. First we compute $\Sigma(z)$ and $W(z)$. Then we compute the Weiss's self-energy $W_{i[1,\dots]}(z)$. We have

$$W_{M[1,\dots,M-1]}(z) = 0 \tag{238}$$

$$W_{M-1[1,\dots,M-2]}(z) = \frac{t_0^2}{z - \varepsilon_0 - \Sigma(z)} \tag{239}$$

and then

$$W_{i[1,\dots,i-1]}(z) = \frac{t_0^2}{z - \varepsilon_0 - \Sigma(z) - W_{i+1[1,\dots,i]}(z)} \tag{240}$$

until $W_{2[1]}(z)$. We can check that

$$W_1(z) = W(z) = \frac{2t_0^2}{z - \varepsilon_0 - \Sigma(z) - W_{2[1]}(z)} \tag{241}$$

There is a difference, depending if we use the RPE or not to compute $W(z)$. Nevertheless, for $M > 50$, the agreement is excellent.

Then, we compute the corresponding Green's functions

$$G_{ii[1,\dots,i-1]}(z) = \frac{1}{z - \varepsilon_0 - \Sigma(z) - W_{i[1,\dots,i-1]}(z)} \tag{242}$$

Then we compute $G_{ij}(z)$ with

$$G_{i,j+1}(z) = -t_0 G_{i,j}(z) G_{j+1,j+1[1,\dots,j]}(z) \tag{243}$$

Then, we compute

$$G(k, z) = \sum_{j=1}^{M} e^{i2\pi kj/M} \, G_{1,j}(z) \qquad (244)$$

But it is safer to choose M odd and to compute

$$G(k, z) = G_{o,o}(z) + \sum_{j=1}^{(M+1)/2} 2cos(2\pi kn/M) \, G_{o,n}(z) \qquad (245)$$

For $M = 11$ one obtains a good agreement and for $M = 21$ we obtain a very good agreement. For $M = 49$, we obtain a very good agreement for all the values of k. We find an excellent agreement for both $W(z)$.

Let us now consider the small polaron problem. We write the program `displlp.cc` to save the first 11 spectral functions. First we compute $W(z)$ and $\Sigma(z)$, next we compute $W_{i[1,...,i-1]}(z)$. Next, we compute the operators

$$\mathbf{G}_{ii[i,...,i-1]}(z) = \frac{1}{z - \varepsilon_0 - \omega_0 b^\dagger b + g_0\omega_0 X - \mathbf{W}_{ii[1,...,i-1]}(z)} \qquad (246)$$

with the diagonal operator

$$\begin{aligned}
\mathbf{W}^{0,0}_{ii[1,...,i-1]}(z) &= W_{i[1,...,i-1]}(z) \\
\mathbf{W}^{n,n}_{ii[1,...,i-1]}(z) &= W(z - n\omega_0)
\end{aligned} \qquad (247)$$

First, we compute $\tilde{G}_{1,1}(z)$

$$\tilde{G}_{1,1}(z) = \sum_m \sum_n U^{m,0}\mathbf{G}^{m,n}_{1,1}(z)U^{n,0} \qquad (248)$$

Next, we compute $\tilde{G}_{1,2}(z)$

$$\tilde{G}_{1,2}(z) = -t_0 \sum_m \sum_n U^{m,0}\mathbf{G}^{m,0}_{11}(z)\mathbf{G}^{0,n}_{22[1]}(z)U^{n,0} \qquad (249)$$

Next, for the other Green's functions, we compute $\tilde{G}_{1,n}(z)$ in three steps and finally we compute

$$\tilde{G}(k, z) = \tilde{G}_{o,o}(z) + \sum_{j=1}^{(M+1)/2} 2cos(2\pi kn/M) \, \tilde{G}_{o,n}(z) \qquad (250)$$

We obtain well defined spectral functions.

We check that the local Green's function corresponds to

$$\tilde{G}(z) = \frac{1}{M} \sum_{\mathbf{k}} \tilde{G}(\mathbf{k}, z) \tag{251}$$

The agreement is perfect.

Dressing

First, we write a program `exact2p.cc` to compute the Green's function $\tilde{G}_{1,1}^{00,00}(z)$ and $\tilde{G}_{1,2}^{00,00}(z)$. We do not compute explicitly the unitary operators U_1 and U_2. Instead we diagonalize first H_{11} and H_{22} and make the change of basis. We use $\varepsilon_0 = 0$, $t_0 = 1$, $\omega_0 = 1$, and $g_0 = 1$, with $P = 25$. We obtain

$$
\begin{array}{cc}
E_\lambda & \mathcal{Z}_{11}(+)\mathcal{Z}_{12}(-) \\
-1.688478504250869650 & 0.430149466868875008 \\
-1.141057011911970731 & \pm 0.375749528331815019 \\
-0.688478504250869539 & 0.000000000000000000 \\
-0.408616163302891688 & 0.041573512920735002 \\
-0.141057011911970898 & 0.000000000000000000 \\
0.292509368378517098 & \pm 0.013460519168691065 \\
0.311521495749130350 & 0.000000000000000000 \\
0.591383836697108367 & 0.000000000000000000 \\
0.846460693436160927 & \pm 0.103999383495229789
\end{array}
\tag{252}
$$

We write the program `white.cc` to diagonalize the two-site Hamiltonian. The first eigenstate with $E_0 = -1.688$ corresponds to the symmetry $k = 0$ and the next with $E_1 = -1.141$, corresponds to the symmetry $k = \pi$, so we construct the reduced matrix for site 1. We have

$$|p\rangle = \sum_{j,m_1,m_2} \Psi_{j,m_1,m_2}(p)\, c_j^\dagger |m_1, m_2\rangle \tag{253}$$

so we get

$$\rho^{m,n}(p) = \sum_{m2} \Psi^*_{1,m,m_2}(p)\Psi_{1,n,m_2}(p) \tag{254}$$

Then, we diagonalize $\rho(p)$ and get $U(p)$. Next, we compute the Green's functions for the electron

$$G_{1,1}^{m,n}(z) = \langle m,0|c_1 \frac{1}{z-H} c_1^\dagger |n,0\rangle \tag{255}$$

and

$$G_{1,2}^{m,n}(z) = \langle m,0|c_1 \frac{1}{z-H} c_2^\dagger |0,n\rangle \tag{256}$$

We use $U(0)$ to compute the Green's functions for the polaron $\tilde{G}_{1,1}^{00,00}(z)$ and $\tilde{G}_{1,2}^{00,00}(z)$.

$$\tilde{G}_{i,j}^{00,00}(z) = \sum_{m,n} U^{m,0} G_{i,j}^{m,n}(z) U^{n,0} \tag{257}$$

We obtain

$$
\begin{array}{cc}
E_p & \mathcal{Z}_{11}(+)\mathcal{Z}_{12}(-) \\
-1.688478425012344353 & 0.460647635452044235 \\
-1.141055651293304329 & \pm 0.282474848867404815 \\
-0.688475201619267341 & 0.018595822843616775 \\
-0.408596071569709784 & 0.006845356931784155 \\
-0.141008612432302949 & \pm 0.008294678957228995 \\
0.292658660920924185 & \pm 0.036577429670409410 \\
0.311575967447863256 & 0.001150416844366987 \\
0.591944962114656859 & 0.000000951634129685 \\
0.846530736410777362 & \pm 0.157090480167243474
\end{array}
\tag{258}
$$

We see that $\tilde{G}(k=0,z)$ has a main pic with $\mathcal{Z}=0.92$ while $\tilde{G}(k=\pi,z)$ has a main pic with $\mathcal{Z}=0.56$.

Next, we do the same with $U(\pi)$. We get

$$
\begin{array}{cc}
E_p & \mathcal{Z}_{11}(+)\mathcal{Z}_{12}(-) \\
-1.688478425012344353 & 0.352163187967980962 \\
-1.141055651293304329 & \pm 0.418408392183039823 \\
-0.688475201619267341 & 0.014669819964840496 \\
-0.408596071569709784 & 0.084227480574715297 \\
-0.141008612432302949 & \pm 0.004504170442218881 \\
0.292658660920924185 & \pm 0.002784171399195930 \\
0.311575967447863256 & 0.001827250565791005 \\
0.591944962114656859 & 0.000209345750042924 \\
0.846530736410777362 & \pm 0.053513502105455815
\end{array}
\tag{259}
$$

This time, $\tilde{G}(k = 0, z)$ has a main pic with $\mathcal{Z} = 0.70$ while $\tilde{G}(k = \pi, z)$ has a main pic with $\mathcal{Z} = 0.84$.

These results should be compared with those of the atomic polaron, that is $\tilde{G}(k = 0, z)$ has a main pic with $\mathcal{Z} = 0.86$ while $\tilde{G}(k = \pi, z)$ has a main pic with $\mathcal{Z} = 0.75$.

Next, we use the program `dmft2.cc` to compute $W(z)$, then use the program `fitAg.cc` to find its poles

$$W(z) = \sum_{k=1}^{N_w} \frac{v_k^2}{z - \varepsilon_k} \tag{260}$$

We obtain 14 poles.

Next, we write the program `whited.cc` to diagonalize the DMFT Hamiltonian

$$H = \varepsilon_0 c^\dagger c + \omega_0 b^\dagger b - g_0 \omega_0 X c^\dagger c + \sum_k \varepsilon_k c_k^\dagger c_k + \sum_k v_k [c^\dagger c_k + c_k^\dagger c] \tag{261}$$

We check that the spectral function of the electron agree with the DMFT result. We find a very good agreement, nearly perfect.

Then we compute the reduced density matrix and diagonalize it to get U and save the matrix $U^{n,0}$.

We modify `dmft2.cc` into `dmft2u.cc` in order to compute $\tilde{G}_{i,j}(z)$ with this operator. We obtain a better result for $k = 0$ and a bad one for $k = \pi$.

Programs

berciu.cc

```
//================================================================//
// berciu.cc
//================================================================//
#include <cmath>
#include <cstdio>
#include <cstdlib>
//================================================================//
const double e0=0.0;
const double w0=1.0;
const double t0=1.0;
```

```cpp
const double g0=1.0;
const double pi=acos(-1.0);
//=========================================================================//
const int M=51;
const int dimB=10;
//=========================================================================//
const double wbeg=-10.0;
const double wend=10.0;
const int wpoints=2000;
const double wpas=(wend-wbeg)/wpoints;
const double eta=0.1;
//=========================================================================//
double wtab[wpoints],As[wpoints],Ag[wpoints],Agk0[wpoints];
double W0[2][wpoints],S[2][wpoints];
//=========================================================================//
int Inear(double a)
{
double a1=floor(a);
double a2=ceil(a);
if(fabs(a1-a)<fabs(a2-a)) return int(a1);
else return int(a2);
}
//=========================================================================//
void mkW0()
{
for(int iw=0;iw<wpoints;iw++)
{
double sumr=0.0;
double sumi=0.0;
for(int k=0;k<M;k++)
{
double ek=e0-2.0*t0*cos(2.0*pi*k/double(M));
double tampr=wtab[iw]-ek;
double tampi=eta;
double denom=tampr*tampr+tampi*tampi;
sumr+=tampr/denom;
sumi-=tampi/denom;
}//endor k
sumr=sumr/double(M);
sumi=sumi/double(M);
double denom=sumr*sumr+sumi*sumi;
W0[0][iw]=wtab[iw]-e0-sumr/denom;
W0[1][iw]=eta+sumi/denom;
}//endfor iw
}
//=========================================================================//
void saveAw()
{
FILE *fp;
fp=fopen("Aw0.dat","w");
for(int iw=0;iw<wpoints;iw++)
{
```

```cpp
double tamp=-2.0*W0[1][iw];
fprintf(fp,"%18.18f_%18.18f_%18.18f\n",wtab[iw],W0[0][iw],tamp);
}
fclose(fp);
}
//=================================================================//
void saveAs()
{
FILE *fp;
fp=fopen("As.dat","w");
for(int iw=0;iw<wpoints;iw++)
{
fprintf(fp,"%18.18f_%18.18f\n",wtab[iw],As[iw]);
}
fclose(fp);
}
//=================================================================//
void saveAg()
{
FILE *fp;
fp=fopen("Ag.dat","w");
for(int iw=0;iw<wpoints;iw++)
{
fprintf(fp,"%18.18f_%18.18f\n",wtab[iw],Ag[iw]);
}
fclose(fp);
}
//=================================================================//
void saveAgk0()
{
FILE *fp;
fp=fopen("Agk0.dat","w");
for(int iw=0;iw<wpoints;iw++)
{
fprintf(fp,"%18.18f_%18.18f\n",wtab[iw],Agk0[iw]);
}
fclose(fp);
}
//=================================================================//
void mkS()
{
for(int iw=0;iw<wpoints;iw++)
{
double sr=0.0;
double si=0.0;
for(int n=dimB;n>0;n--)
{
int i2=Inear(iw*1.0-n*w0/wpas);
double tampr=wtab[iw]-e0-n*w0-sr;
double tampi=eta-si;
if((i2>=0)&&(i2<wpoints)){ tampr-=W0[0][i2]; tampi-=W0[1][i2];}
double denom=tampr*tampr+tampi*tampi;
```

```cpp
sr=n*g0*g0*w0*w0*tampr/denom;
si=-n*g0*g0*w0*w0*tampi/denom;
}//end n
S[0][iw]=sr;
S[1][iw]=si;
As[iw]=-2.0*si;
double tampr=wtab[iw]-e0+2.0*t0-sr;
double tampi=eta-si;
Agk0[iw]=2.0*tampi/(tampr*tampr+tampi*tampi);
}//end iw
}
//================================================================//
void mkAloc()
{
for(int iw=0;iw<wpoints;iw++)
{
double sumr=0.0;
double sumi=0.0;
for(int k=0;k<M;k++)
{
double ek=e0-2.0*t0*cos(2.0*pi*k/double(M));
double tampr=wtab[iw]-ek-S[0][iw];
double tampi=eta-S[1][iw];
double denom=tampr*tampr+tampi*tampi;
sumr+=tampr/denom;
sumi-=tampi/denom;
}//endor k
Ag[iw]=-2.0*sumi/double(M);
}
}
//================================================================//
main()
{
for(int iw=0;iw<wpoints;iw++) wtab[iw]=wbeg+iw*wpas;
mkW0();
saveAw();
mkS();
mkAloc();
saveAs();
saveAg();
saveAgk0();
}
//================================================================//
```

dmft.cc

```cpp
//================================================================//
// dmft.cc
//================================================================//
#include <cmath>
#include <cstdio>
#include <cstdlib>
```

```cpp
#include "ludcmp.h"
#include "lubksb.h"
#include "mprove.h"
//==============================================================//
const double e0=0.0;
const double w0=1.0;
const double t0 =1.0;
const double g0=1.0;
const double pi=acos(-1.0);
//==============================================================//
const int M=51;
const int dimB=25;
//==============================================================//
const double wbeg=-10.0;
const double wend=10.0;
const int wpoints=5000;
const double wpas=(wend-wbeg)/wpoints;
const double eta=0.1;
//==============================================================//
int iz;
double Erreur;
double wtab[wpoints],As[wpoints],Ag[wpoints],Agk0[wpoints];
double Apol[wpoints];
double W[2][wpoints],S[2][wpoints],Sold[2][wpoints];
double Kr[dimB][dimB],Ki[dimB][dimB];
double Gr[dimB][dimB],Gi[dimB][dimB];
double U[dimB];
//==============================================================//
int Inear(double a)
{
double a1=floor(a);
double a2=ceil(a);
if(fabs(a1-a)<fabs(a2-a)) return int(a1);
else return int(a2);
}
//==============================================================//
void initS()
{
for(int iw=0;iw<wpoints;iw++)
{
Sold[0][iw]=0.0;
Sold[1][iw]=0.0;
}
}
//==============================================================//
void mkW()
{
for(int iw=0;iw<wpoints;iw++)
{
double sumr=0.0;
double sumi=0.0;
for(int k=0;k<M;k++)
```

```cpp
{
double  ek=e0 −2.0∗t0∗cos (2.0∗ pi∗k / double (M));
double  tampr=wtab [iw]−ek−Sold [0][ iw ];
double  tampi=eta −Sold [1][ iw ];
double  denom=tampr∗tampr+tampi∗tampi ;
sumr+=tampr / denom ;
sumi −=tampi / denom ;
} // endor  k
sumr=sumr / double (M);
sumi=sumi / double (M);
double  denom=sumr∗sumr+sumi∗sumi ;
W[0][ iw ]= wtab [ iw]−e0−Sold [0][ iw]− sumr / denom ;
W[1][ iw ]= eta −Sold [1][ iw ]+ sumi / denom ;
} // endfor  iw
}
//=================================================================//
void  saveAw ()
{
FILE  ∗fp ;
fp=fopen ("Aw0. dat" ,"w" );
for ( int  iw =0;iw<wpoints ;iw ++)
{
double  tamp=−2.0∗W[1][ iw ];
fprintf (fp ,"%18.18 f _%18.18 f _%18.18 f \n" ,wtab [ iw ],W[0][ iw ] ,tamp );
}
fclose ( fp );
}
//=================================================================//
void  saveAs ()
{
FILE  ∗fp ;
fp=fopen ("As. dat" ,"w" );
for ( int  iw =0;iw<wpoints ;iw ++)
{
fprintf (fp ,"%18.18 f _%18.18 f \n" ,wtab [ iw ] ,As[ iw ] );
}
fclose ( fp );
}
//=================================================================//
void  saveAg ()
{
FILE  ∗fp ;
fp=fopen ("Ag. dat" ,"w" );
for ( int  iw =0;iw<wpoints ;iw ++)
{
fprintf (fp ,"%18.18 f _%18.18 f \n" ,wtab [ iw ] ,Ag[ iw ] );
}
fclose ( fp );
}
//=================================================================//
void  saveAgk0 ()
{
```

```cpp
FILE *fp;
fp=fopen("Agk0.dat","w");
for(int iw=0;iw<wpoints;iw++)
{
fprintf(fp,"%18.18f_%18.18f\n",wtab[iw],Agk0[iw]);
}
fclose(fp);
}
//=================================================================//
void saveApol()
{
FILE *fp;
fp=fopen("Apol.dat","w");
for(int iw=0;iw<wpoints;iw++)
{
fprintf(fp,"%18.18f_%18.18f\n",wtab[iw],Apol[iw]);
}
fclose(fp);
}
//=================================================================//

void mkS()
{
for(int iw=0;iw<wpoints;iw++)
{
double sr=0.0;
double si=0.0;
for(int n=dimB;n>0;n--)
{
int i2=Inear(iw*1.0-n*w0/wpas);
double tampr=wtab[iw]-e0-n*w0-sr;
double tampi=eta-si;
if((i2>=0)&&(i2<wpoints)){ tampr-=W[0][i2]; tampi-=W[1][i2];}
double denom=tampr*tampr+tampi*tampi;
sr=n*g0*g0*w0*w0*tampr/denom;
si=-n*g0*g0*w0*w0*tampi/denom;
}//end n
S[0][iw]=sr;
S[1][iw]=si;
As[iw]=-2.0*si;
double tampr=wtab[iw]-e0-sr-W[0][iw];
double tampi=eta-si-W[1][iw];
Ag[iw]=2.0*tampi/(tampr*tampr+tampi*tampi);
tampr=wtab[iw]-e0+2.0*t0-sr;
tampi=eta-si;
Agk0[iw]=2.0*tampi/(tampr*tampr+tampi*tampi);

}//end iw
}
//=================================================================//
void mkError()
{
```

```cpp
Erreur =0.0;
for ( int  iw =0;iw<wpoints ;iw++)
{
Erreur+=fabs (S[0][iw]− Sold [0][iw])+ fabs (S[1][iw]− Sold [1][iw]);
Sold [0][iw]=0.5∗ Sold [0][iw]+0.5∗S[0][iw];
Sold [1][iw]=0.5∗ Sold [1][iw]+0.5∗S[1][iw];
}
}
//=================================================================//
void mkKloc ()
{
for ( int  i =0;i<dimB ; i ++)  for ( int  j =0;j<dimB ; j ++)  Kr[ i ][ j ]=0.0;
for ( int  i =0;i<dimB ; i ++)  for ( int  j =0;j<dimB ; j ++)  Ki[ i ][ j ]=0.0;
for ( int  n=0;n<dimB ; n++)  Kr[ n ][ n ]=wtab [ iz ]−e0−n∗w0;
for ( int  n=0;n<dimB ; n++)  Ki[ n ][ n ]=eta ;
for ( int  i =1;i<dimB ; i ++)  Kr[ i −1][ i ]=Kr[ i ][ i −1]=g0 ∗w0∗ sqrt (1.0∗ i );
for ( int  n=0;n<dimB ; n++)
{
int  i2 =Inear ( iz ∗1.0−n∗w0/ wpas );
if ( i2 >=0){Kr[ n ][ n ]−=W[0][ i2 ]; Ki[ n ][ n ]−=W[1][ i2 ];}
}
}
//=================================================================//
void mkGloc ()
{
double ∗∗g_mat ; g_mat=new double ∗[2∗dimB ];
for ( int  i =0;i<2∗dimB ; i ++)  g_mat [ i ]=new double [2∗dimB ];
double ∗∗glu_mat ; glu_mat=new double ∗[2∗dimB ];
for ( int  i =0;i<2∗dimB ; i ++)  glu_mat [ i ]=new double [2∗dimB ];
double ∗x_vec=new double [2∗dimB ];
double ∗b_vec=new double [2∗dimB ];
int ∗perm_vec=new int [2∗dimB ];
double dperm ;
double err ;

for ( int  i =0;i<dimB ; i ++)
{
for ( int  j =0;j<dimB ; j ++)
{
g_mat [ i ][ j ]=Kr[ i ][ j ];
g_mat [ i ][ j+dimB]=− Ki[ i ][ j ];
g_mat [ i+dimB ][ j ]=Ki[ i ][ j ];
g_mat [ i+dimB ][ j+dimB ]=Kr[ i ][ j ];
}
}

for ( int  i =0;i<2∗dimB ; i ++)  for ( int  j =0;j<2∗dimB ; j ++)
glu_mat [ i ][ j ]=g_mat [ i ][ j ];

ludcmp ( glu_mat ,2∗dimB , perm_vec , dperm );

for ( int  n=0;n<dimB ; n++)
```

```cpp
{
for(int  i=0;i<2*dimB;i++)  x_vec[i]=0.0;x_vec[n]=1.0;
lubksb(glu_mat,2*dimB,perm_vec,x_vec);
for(int  i=0;i<2*dimB;i++)  b_vec[i]=0.0;b_vec[n]=1.0;
do
{
mprove(g_mat,glu_mat,2*dimB,perm_vec,b_vec,x_vec,err);
}while(err>10e-12);
for(int  i=0;i<dimB;i++)  Gr[i][n]=x_vec[i];
for(int  i=0;i<dimB;i++)  Gi[i][n]=x_vec[i+dimB];
}

for(int  i=0;i<2*dimB;i++)  delete[]  g_mat[i];delete[]  g_mat;
for(int  i=0;i<2*dimB;i++)  delete[]  glu_mat[i];delete[]  glu_mat;
delete[]  x_vec;
delete[]  b_vec;
delete[]  perm_vec;
}
//================================================================//
void mkU()
{
double  tamp=exp(-0.5*g0*g0);
U[0]=tamp;
for(int  n=1;n<dimB;n++)
{
tamp=tamp*g0/sqrt(1.0*n);
U[n]=tamp;
}
}
//================================================================//
void mkApol()
{
double  tamp=0.0;
for(int  i=0;i<dimB;i++)
{
for(int  j=0;j<dimB;j++)
{
tamp+=U[i]*U[j]*Gi[i][j];
}
}
Apol[iz]=-2.0*tamp;
}
//================================================================//
main()
{
for(int  iw=0;iw<wpoints;iw++)  wtab[iw]=wbeg+iw*wpas;
initS();
do
{
mkW();
saveAw();
mkS();
```

```
saveAs ();
saveAg ();
saveAgk0 ();
mkError ();
printf ("Erreur=%18.18f\n", Erreur );
} while ( Erreur >10e −12);
mkU ();
for ( iz =0; iz <wpoints ; iz ++)
{
mkKloc ();
mkGloc ();
mkApol ();
}
saveApol ();
}
//================================================================//
```

exact.cc

```
//================================================================//
// exact . cc
// reduced    basis  no  crossing
//================================================================//
#include <cmath>
#include <cstdio>
#include <cstdlib>
#include "jacobi .h"
#include "eigsrt .h"
//================================================================//
const double e0 =0.0;
const double w0 =1.0;
const double t0 =1.0;
const double g0 =1.0;
//================================================================//
const int M=10;
const int dimB =10;
const int dimBB=M*( dimB −1)+1;
const int dimH=M*dimBB ;
//================================================================//
int NB[ dimBB ][ M];
double V[M][ dimBB ][ dimBB ];
double U[ dimBB ];
double H[ dimH ][ dimH ];
double Ep[ dimH ], V11e[ dimH ], V11p[ dimH ];
//================================================================//
void mkNB ()
{
for ( int i =0; i <dimBB ; i ++) for ( int j =0; j <M; j ++) NB[ i ][ j ]=0;
for ( int i =0; i <dimB ; i ++) NB[ i ][0]= i ;
for ( int j =1; j <M; j ++)
{
for ( int i =1; i <dimB ; i ++) NB[ j *( dimB −1)+i ][ j ]= i ;
```

```cpp
}
/*
for(int  i=0;i<dimBB;i++)
{
printf("%d  ",i);
for(int  j=0;j<M;j++)  printf("%d  ",NB[i][j]);
printf("\n");
}
*/
}
//=======================================================================//
void mkV()
{
for(int  n=0;n<M;n++)
for(int  i=0;i<dimBB;i++)  for(int  j=0;j<dimBB;j++)  V[n][i][j]=0.0;
for(int  i=1;i<dimB;i++)  V[0][i-1][i]=V[0][i][i-1]=sqrt(1.0*i);
for(int  n=1;n<M;n++)
{
for(int  i=0;i<dimB;i++)  for(int  j=0;j<dimB;j++)
if((((i==0)&&(j==1))||((i==1)&&(j==0)))
V[n][0][n*(dimB-1)+1]=V[n][n*(dimB-1)+1][0]=1.0;
else
V[n][i+n*(dimB-1)][j+n*(dimB-1)]=V[0][i][j];
}
}
//=======================================================================//
void seeV()
{
for(int  n=0;n<M;n++)
{
printf("n=%d\n",n);
for(int  i=0;i<dimBB;i++)
{
for(int  j=0;j<dimBB;j++)  printf("%3.2f_",V[n][i][j]);
printf("\n");
}
}
}//=====================================================================//
void mkD()
{
for(int  n=0;n<M;n++)
for(int  i=0;i<dimBB;i++)  for(int  j=0;j<dimBB;j++)
V[n][i][j]=-g0*w0*V[n][i][j];

for(int  n=0;n<M;n++)
{
for(int  i=0;i<dimBB;i++)
{
int  inb=0;
for(int  j=0;j<M;j++)  inb+=NB[i][j];
V[n][i][i]+=e0+inb*w0;
}
```

```cpp
}
}
//=================================================================//
void mkHD()
{
for(int  i=0;i<dimH;i++)  for(int  j=0;j<dimH;j++)  H[i][j]=0.0;
for(int  n=0;n<M;n++)
{
for(int  i=0;i<dimBB;i++)  for(int  j=0;j<dimBB;j++)
H[i+n*dimBB][j+n*dimBB]=V[n][i][j];
}
}
//=================================================================//
void mkHT()
{
for(int  n=1;n<M;n++)
{
for(int  i=0;i<dimBB;i++)
H[n*dimBB+i][n*dimBB-dimBB+i]=H[n*dimBB-dimBB+i][n*dimBB+i]=-t0;
}
for(int  i=0;i<dimBB;i++)
H[M*dimBB-dimBB+i][i]=H[i][M*dimBB-dimBB+i]=-t0;
}
//=================================================================//
void seeH()
{
printf("H=\n");
for(int  i=0;i<dimH;i++)
{
for(int  j=0;j<dimH;j++)  printf("%3.2f_",H[i][j]);
printf("\n");
}
}
//=================================================================//
void diagH()
{
double  **h_mat;h_mat=new double  *[dimH];
for(int  i=0;i<dimH;i++)  h_mat[i]=new double[dimH];
double  **v_mat;v_mat=new double  *[dimH];
for(int  i=0;i<dimH;i++)  v_mat[i]=new double[dimH];
double  *d_vec=new double[dimH];

for(int  i=0;i<dimH;i++)  for(int  j=0;j<dimH;j++)
                          h_mat[i][j]=H[i][j];
jacobi(h_mat,dimH,d_vec,v_mat,10e-24);
eigsrt(d_vec,v_mat,dimH);
for(int  i=0;i<dimH;i++)  Ep[i]=d_vec[i];
for(int  i=0;i<dimH;i++)  V11e[i]=v_mat[0][i]*v_mat[0][i];
for(int  i=0;i<dimH;i++)
{
V11p[i]=0.0;
for(int  n=0;n<dimBB;n++)  for(int  m=0;m<dimBB;m++)
```

```cpp
V11p[i]+=v_mat[n][i]*U[n]*U[m]*v_mat[m][i];
}

for(int i=0;i<dimH;i++) delete[] h_mat[i]; delete[] h_mat;
for(int i=0;i<dimH;i++) delete[] v_mat[i]; delete[] v_mat;
delete[] d_vec;
}
//================================================================//
void mkU()
{
double tamp=exp(-0.5*g0*g0);
U[0]=tamp;
for(int n=1;n<dimBB;n++)
{
tamp=tamp*g0/sqrt(1.0*n);
U[n]=tamp;
}
}
//================================================================//
void saveG11()
{
FILE *fp;
fp=fopen("ev11e.dat","w");
for(int i=0;i<dimH;i++)
{
fprintf(fp,"%18.18f_%18.18f\n",Ep[i],V11e[i]);
}
fclose(fp);
}
//================================================================//
void saveP11()
{
FILE *fp;
fp=fopen("ev11p.dat","w");
for(int i=0;i<dimH;i++)
{
fprintf(fp,"%18.18f_%18.18f\n",Ep[i],V11p[i]);
}
fclose(fp);
}
//================================================================//
main()
{
printf("M=%d_P=%d_dimH=%d\n",M,dimB,dimH);
mkNB();
mkV();
//seeV();
mkD();
//seeV();
mkHD();
mkHT();
//seeH();
```

```cpp
mkU();
diagH();
printf("E0=%18.18f\n",Ep[0]);
saveG11();
saveP11();
}
//==================================================================//
```

berciu2.cc

```cpp
//==================================================================//
//berciu2.cc
//we compute W(z) via RPE
//==================================================================//
#include <cmath>
#include <cstdio>
#include <cstdlib>
//==================================================================//
const double e0=0.0;
const double w0=1.0;
const double t0=1.0;
const double g0=1.0;
const double pi=acos(-1.0);
//==================================================================//
const int M=51;
const int dimB=10;
//==================================================================//
const double wbeg=-10.0;
const double wend=10.0;
const int wpoints=2000;
const double wpas=(wend-wbeg)/wpoints;
const double eta=0.1;
//==================================================================//
double wtab[wpoints],As[wpoints],Ag[wpoints],Agk0[wpoints];
double W0[2][wpoints],S[2][wpoints];
//==================================================================//
int Inear(double a)
{
double a1=floor(a);
double a2=ceil(a);
if(fabs(a1-a)<fabs(a2-a)) return int(a1);
else return int(a2);
}
//==================================================================//
void mkW0()
{
for(int iw=0;iw<wpoints;iw++)
{
double sumr=0.0;
double sumi=0.0;
for(int n=M-2;n>0;n--)
{
```

```cpp
    double tampr=wtab[iw]-e0-sumr;
    double tampi=eta-sumi;
    double denom=tampr*tampr+tampi*tampi;
    sumr=t0*t0*tampr/denom;
    sumi=-t0*t0*tampi/denom;
    }//endor n
    double tampr=wtab[iw]-e0-sumr;
    double tampi=eta-sumi;
    double denom=tampr*tampr+tampi*tampi;
    W0[0][iw]=2.0*t0*t0*tampr/denom;
    W0[1][iw]=-2.0*t0*t0*tampi/denom;
    }//endfor iw
}
//================================================================//
void saveAw()
{
FILE *fp;
fp=fopen("Aw0.dat","w");
for(int iw=0;iw<wpoints;iw++)
{
double tamp=-2.0*W0[1][iw];
fprintf(fp,"%18.18f_%18.18f_%18.18f\n",wtab[iw],W0[0][iw],tamp);
}
fclose(fp);
}
//================================================================//
void saveAs()
{
FILE *fp;
fp=fopen("As.dat","w");
for(int iw=0;iw<wpoints;iw++)
{
fprintf(fp,"%18.18f_%18.18f\n",wtab[iw],As[iw]);
}
fclose(fp);
}
//================================================================//
void saveAg()
{
FILE *fp;
fp=fopen("Ag.dat","w");
for(int iw=0;iw<wpoints;iw++)
{
fprintf(fp,"%18.18f_%18.18f\n",wtab[iw],Ag[iw]);
}
fclose(fp);
}
//================================================================//
void saveAgk0()
{
FILE *fp;
fp=fopen("Agk0.dat","w");
```

```cpp
for ( int  iw =0; iw<wpoints ; iw ++)
{
fprintf ( fp , "%18.18 f _%18.18 f \n" , wtab [ iw ] , Agk0 [ iw ] );
}
fclose ( fp );
}
//=============================================================//
void  mkS ()
{
for ( int  iw =0; iw<wpoints ; iw ++)
{
double  sr =0.0;
double  si =0.0;
for ( int  n=dimB ; n>0; n--)
{
int  i2=Inear ( iw *1.0 - n*w0/ wpas );
double  tampr=wtab [ iw ]-e0-n*w0-sr ;
double  tampi=eta -si ;
if (( i2 >=0)&&( i2<wpoints )){ tampr -=W0[ 0 ][ i2 ] ; tampi -=W0[ 1 ][ i2 ] ;}
double  denom=tampr *tampr +tampi *tampi ;
sr=n*g0 *g0 *w0 *w0 *tampr / denom ;
si=-n*g0 *g0 *w0 *w0 *tampi / denom ;
}// end  n
S [ 0 ][ iw ]= sr ;
S [ 1 ][ iw ]= si ;
As [ iw ]=-2.0* si ;
double  tampr=wtab [ iw ]-e0 +2.0* t0 -sr ;
double  tampi=eta -si ;
Agk0 [ iw ]=2.0* tampi /( tampr *tampr +tampi *tampi );
}// end  iw
}
//=============================================================//
void  mkAloc ()
{
for ( int  iw =0; iw<wpoints ; iw ++)
{
double  sumr =0.0;
double  sumi =0.0;
for ( int  k =0; k<M; k ++)
{
double  ek=e0 -2.0* t0 *cos (2.0* pi *k/ double (M));
double  tampr=wtab [ iw ]-ek-S [ 0 ][ iw ] ;
double  tampi=eta -S [ 1 ][ iw ] ;
double  denom=tampr *tampr +tampi *tampi ;
sumr+=tampr / denom ;
sumi -=tampi / denom ;
}// endor  k
Ag [ iw ]=-2.0* sumi / double (M);
}
}
//=============================================================//
main ()
```

```cpp
{
for(int iw=0;iw<wpoints;iw++) wtab[iw]=wbeg+iw*wpas;
mkW0();
saveAw();
mkS();
mkAloc();
saveAs();
saveAg();
saveAgk0();
}
//=====================================================================//
```

local.cc

```cpp
//=====================================================================//
//local.cc
//local Green's function with W0(z)
//=====================================================================//
#include <cmath>
#include <cstdio>
#include <cstdlib>
#include "ludcmp.h"
#include "lubksb.h"
#include "mprove.h"
//=====================================================================//
const double e0=0.0;
const double w0=1.0;
const double t0=1.0;
const double g0=1.0;
const double pi=acos(-1.0);
//=====================================================================//
const int M=10;
const int dimB=10;
//=====================================================================//
const double wbeg=-10.0;
const double wend=10.0;
const int wpoints=2000;
const double wpas=(wend-wbeg)/wpoints;
const double eta=0.1;
//=====================================================================//
int iz;
double wtab[wpoints];
double W0[2][wpoints],W1[2][wpoints],S0[2][wpoints];
double G00[wpoints],P00[wpoints];
double U[dimB];
double Kr[dimB][dimB],Ki[dimB][dimB];
double Gr[dimB][dimB],Gi[dimB][dimB];
//=====================================================================//
int Inear(double a)
{
double a1=floor(a);
double a2=ceil(a);
```

```cpp
if ( fabs ( a1−a)<fabs ( a2−a )) return int ( a1 );
else return int ( a2 );
}
//===============================================================//
void mkU ( )
{
double tamp=exp ( −0.5 ∗ g0 ∗ g0 );
U[0]=tamp ;
for ( int n=1;n<dimB ; n++)
{
tamp=tamp ∗ g0 / sqrt (1.0 ∗ n );
U[ n]=tamp ;
}
}
//===============================================================//
void mkW0 ( )
{
for ( int iw =0;iw<wpoints ;iw++)
{
double sumr =0.0;
double sumi =0.0;
for ( int k=0;k<M; k++)
{
double ek=e0 −2.0 ∗ t0 ∗ cos (2.0 ∗ pi ∗ k / double (M));
double tampr=wtab [ iw]−ek ;
double tampi=eta ;
double denom=tampr ∗ tampr+tampi ∗ tampi ;
sumr+=tampr / denom ;
sumi−=tampi / denom ;
} // endor k
sumr=sumr / double (M);
sumi=sumi / double (M);
double denom=sumr ∗ sumr+sumi ∗ sumi ;
W0[0] [ iw]=wtab [ iw]−e0−sumr / denom ;
W0[1] [ iw]=eta+sumi / denom ;
} // endfor iw
}
//===============================================================//
void mkW1 ( )
{
for ( int iw =0;iw<wpoints ;iw++)
{
double sumr =0.0;
double sumi =0.0;
for ( int k=0;k<M; k++)
{
double ek=e0 −2.0 ∗ t0 ∗ cos (2.0 ∗ pi ∗ k / double (M));
double tampr=wtab [ iw]−ek−S0 [0] [ iw ];
double tampi=eta −S0 [1] [ iw ];
double denom=tampr ∗ tampr+tampi ∗ tampi ;
sumr+=tampr / denom ;
sumi−=tampi / denom ;
```

```cpp
}//endor k
sumr=sumr/double(M);
sumi=sumi/double(M);
double denom=sumr*sumr+sumi*sumi;
W1[0][iw]=wtab[iw]-e0-S0[0][iw]-sumr/denom;
W1[1][iw]=eta-S0[1][iw]+sumi/denom;
}//endfor iw
}
//=================================================================//
void mkS0()
{
for(int iw=0;iw<wpoints;iw++)
{
double sr=0.0;
double si=0.0;
for(int n=dimB;n>0;n--)
{
int i2=Inear(iw*1.0-n*w0/wpas);
double tampr=wtab[iw]-e0-n*w0-sr;
double tampi=eta-si;
if((i2>=0)&&(i2<wpoints)){tampr-=W0[0][i2];tampi-=W0[1][i2];}
double denom=tampr*tampr+tampi*tampi;
sr=n*g0*g0*w0*w0*tampr/denom;
si=-n*g0*g0*w0*w0*tampi/denom;
}//end n
S0[0][iw]=sr;
S0[1][iw]=si;
}//end iw
}
//=================================================================//
void mkKloc()
{
for(int i=0;i<dimB;i++) for(int j=0;j<dimB;j++) Kr[i][j]=0.0;
for(int i=0;i<dimB;i++) for(int j=0;j<dimB;j++) Ki[i][j]=0.0;
for(int n=0;n<dimB;n++) Kr[n][n]=wtab[iz]-e0-n*w0;
for(int n=0;n<dimB;n++) Ki[n][n]=eta;
for(int i=1;i<dimB;i++) Kr[i-1][i]=Kr[i][i-1]=g0*w0*sqrt(1.0*i);
Kr[0][0]-=W1[0][iz];
Ki[0][0]-=W1[1][iz];
for(int n=1;n<dimB;n++)
{
int i2=Inear(iz*1.0-n*w0/wpas);
if(i2>=0){Kr[n][n]-=W0[0][i2];Ki[n][n]-=W0[1][i2];}
}
}
//=================================================================//
void mkGloc()
{
double **g_mat;g_mat=new double*[2*dimB];
for(int i=0;i<2*dimB;i++) g_mat[i]=new double[2*dimB];
double **glu_mat;glu_mat=new double*[2*dimB];
for(int i=0;i<2*dimB;i++) glu_mat[i]=new double[2*dimB];
```

```cpp
double *x_vec=new double[2*dimB];
double *b_vec=new double[2*dimB];
int *perm_vec=new int[2*dimB];
double dperm;
double err;

for(int i=0;i<dimB;i++)
{
for(int j=0;j<dimB;j++)
{
g_mat[i][j]=Kr[i][j];
g_mat[i][j+dimB]=-Ki[i][j];
g_mat[i+dimB][j]=Ki[i][j];
g_mat[i+dimB][j+dimB]=Kr[i][j];
}
}

for(int i=0;i<2*dimB;i++) for(int j=0;j<2*dimB;j++)
glu_mat[i][j]=g_mat[i][j];

ludcmp(glu_mat,2*dimB,perm_vec,dperm);

for(int n=0;n<dimB;n++)
{
for(int i=0;i<2*dimB;i++) x_vec[i]=0.0;x_vec[n]=1.0;
lubksb(glu_mat,2*dimB,perm_vec,x_vec);
for(int i=0;i<2*dimB;i++) b_vec[i]=0.0;b_vec[n]=1.0;
do
{
mprove(g_mat,glu_mat,2*dimB,perm_vec,b_vec,x_vec,err);
}while(err>10e-12);
for(int i=0;i<dimB;i++) Gr[i][n]=x_vec[i];
for(int i=0;i<dimB;i++) Gi[i][n]=x_vec[i+dimB];
}

for(int i=0;i<2*dimB;i++) delete[] g_mat[i];delete[] g_mat;
for(int i=0;i<2*dimB;i++) delete[] glu_mat[i];delete[] glu_mat;
delete[] x_vec;
delete[] b_vec;
delete[] perm_vec;
}
//================================================================//
void mkP00()
{
double tamp=0.0;
for(int i=0;i<dimB;i++)
{
for(int j=0;j<dimB;j++)
{
tamp+=U[i]*U[j]*Gi[i][j];
}
}
```

```cpp
G00[iz]=-2.0*Gi[0][0];
P00[iz]=-2.0*tamp;
}
//===================================================================//
void saveG00()
{
FILE *fp;
fp=fopen("G00.dat","w");
for(int iw=0;iw<wpoints;iw++)
fprintf(fp,"%18.18f_%18.18f\n",wtab[iw],G00[iw]);
fclose(fp);
}
//===================================================================//
void saveP00()
{
FILE *fp;
fp=fopen("P00.dat","w");
for(int iw=0;iw<wpoints;iw++)
fprintf(fp,"%18.18f_%18.18f\n",wtab[iw],P00[iw]);
fclose(fp);
}
//===================================================================//

main()
{
for(int iw=0;iw<wpoints;iw++)  wtab[iw]=wbeg+iw*wpas;
mkW0();
mkS0();
mkW1();
mkU();
for(iz=0;iz<wpoints;iz++)
{
mkKloc();
mkGloc();
mkP00();
}
saveG00();
saveP00();
}
//===================================================================//
```

exact2.cc

```cpp
//===================================================================//
// exact2.cc
//===================================================================//
// Holstein 2 sites
// Green's function G12(z)
// restricted basis no crossing
//===================================================================//
#include <cmath>
#include <cstdio>
```

```cpp
#include <cstdlib>
#include "jacobi.h"
#include "eigsrt.h"
//===============================================================//
const double e0=0.0;
const double w0=1.0;
const double t0=1.0;
const double g0=1.0;
const int dimB=10;
const int dimBB=2*dimB-1;
const int dimH=2*dimBB;
//===============================================================//
const double wbeg=-10.0;
const double wend=10.0;
const int wpoints=2000;
const double wpas=(wend-wbeg)/wpoints;
const double eta=0.1;
//===============================================================//
int NB1[dimBB],NB2[dimBB],NBT[dimBB];
double U[dimB];
double H1[dimBB][dimBB],H2[dimBB][dimBB];
double G11[dimBB][dimBB][dimH],G12[dimBB][dimBB][dimH];
double P12[dimH],P22[dimH];
double H[dimH][dimH];
double Ep[dimH],Vp[dimH][dimH];
double wtab[wpoints],G11_0000[2][wpoints],G12_0000[2][wpoints];
double P12_0000[2][wpoints],P22_0000[2][wpoints];
//===============================================================//
void mkNB()
{
for(int i=0;i<dimBB;i++) NB1[i]=NB2[i]=0;
for(int i=0;i<dimB;i++) NB1[i]=i;
for(int i=1;i<dimB;i++) NB2[i+dimB-1]=i;
for(int i=0;i<dimBB;i++) NBT[i]=NB1[i]+NB2[i];
//for(int i=0;i<dimBB;i++) printf("%d %d %d\n",i,NB1[i],NB2[i]);
}
//===============================================================//
void mkU()
{
double tamp=exp(-0.5*g0*g0);
U[0]=tamp;
for(int n=1;n<dimB;n++)
{
tamp=tamp*g0/sqrt(1.0*n);
U[n]=tamp;
}
}
//===============================================================//
void mkV1()
{
for(int i=0;i<dimBB;i++) for(int j=0;j<dimBB;j++) H1[i][j]=0.0;
for(int i=1;i<dimB;i++) H1[i-1][i]=H1[i][i-1]=sqrt(1.0*i);
```

```cpp
}
//================================================================//
void mkV2()
{
for(int i=0;i<dimBB;i++) for(int j=0;j<dimBB;j++) H2[i][j]=0.0;
H2[0][dimB]=H2[dimB][0]=1.0;
int pos=dimB-1;
for(int i=2;i<dimB;i++)
H2[pos+i-1][pos+i]=H2[pos+i][pos+i-1]=sqrt(1.0*i);
}
//================================================================//
void mkD1()
{
for(int i=0;i<dimBB;i++) for(int j=0;j<dimBB;j++)
H1[i][j]=-g0*w0*H1[i][j];
for(int i=0;i<dimBB;i++) H1[i][i]+=e0+w0*NBT[i];
}
//================================================================//
void mkD2()
{
for(int i=0;i<dimBB;i++) for(int j=0;j<dimBB;j++)
H2[i][j]=-g0*w0*H2[i][j];
for(int i=0;i<dimBB;i++) H2[i][i]+=e0+w0*NBT[i];
}
//================================================================//
void seeH1()
{
printf("H1\n");
for(int i=0;i<dimBB;i++)
{
for(int j=0;j<dimBB;j++) printf("%3.2f_",H1[i][j]);
printf("\n");
}
}
//================================================================//
void seeH2()
{
printf("H2\n");
for(int i=0;i<dimBB;i++)
{
for(int j=0;j<dimBB;j++) printf("%3.2f_",H2[i][j]);
printf("\n");
}
}
//================================================================//
void mkHD()
{
for(int i=0;i<dimH;i++) for(int j=0;j<dimH;j++) H[i][j]=0.0;
for(int i=0;i<dimBB;i++)
{
for(int j=0;j<dimBB;j++)
{
```

```cpp
H[ i ][ j ]=H1[ i ][ j ];
H[ i+dimBB ][ j+dimBB]=H2[ i ][ j ];
}
}
}
//===============================================================//
void mkHT ()
{
for(int i =0;i<dimBB ; i ++) H[ dimBB+i ][ i ]=H[ i ][ dimBB+i ]=−t0 ;
}
//===============================================================//
void seeH ()
{
printf ("H=\n" );
for(int i =0;i<dimH ; i ++)
{
for(int j =0;j<dimH; j ++) printf ("%3.2 f␣",H[ i ][ j ]);
printf ("\n" );
}
}
//===============================================================//
void diagH ()
{
double **h_mat ; h_mat=new double *[dimH ];
for(int i =0;i<dimH; i ++) h_mat[ i ]=new double[dimH ];
double **v_mat ; v_mat=new double *[dimH ];
for(int i =0;i<dimH; i ++) v_mat[ i ]=new double[dimH ];
double *d_vec=new double[dimH ];

for(int i =0;i<dimH; i ++) for(int j =0;j<dimH; j ++) h_mat[i ][ j ]=H[i ][ j ];
jacobi (h_mat , dimH, d_vec , v_mat ,10 e−24);
eigsrt (d_vec , v_mat , dimH );
for(int i =0;i<dimH; i ++) Ep[ i ]=d_vec [ i ];
for(int i =0;i<dimH; i ++) for(int j =0;j<dimH; j ++)
                          Vp[ i ][ j ]=v_mat [ i ][ j ];

for(int i =0;i<dimH; i ++) delete [] h_mat[ i ]; delete [] h_mat ;
for(int i =0;i<dimH; i ++) delete [] v_mat[ i ]; delete [] v_mat ;
delete [] d_vec ;
}
//===============================================================//
void mkG11 ()
{
int C1[ dimB ];
for(int i =0;i<dimB ; i ++) C1[ i ]=i ;
for(int i =0;i<dimB ; i ++)
{
for(int j =0;j<dimB ; j ++)
{
for(int p=0;p<dimH ; p++)
{
int posi=C1[ i ];
```

```cpp
  int  posj=C1[ j ];
G11[ i ][ j ][ p]=Vp[ posi ][ p]*Vp[ posj ][ p];
}
}
}
}
//===================================================================//
void  mkG12()
{
int  C1[dimB ];
int  C2[dimB ];
for ( int  i =0; i <dimB ; i ++)  C1[ i ]= i ;
C2[0]=0;
for ( int  i =1; i <dimB ; i ++)  C2[ i ]=dimB−1+i ;
//for( int  i =0; i <dimB ; i ++)  printf("%d  %d\n", i , C2[ i ]);
for ( int  i =0; i <dimB ; i ++)
{
for ( int  j =0; j <dimB ; j ++)
{
for ( int  p =0; p<dimH ; p++)
{
int  posi=C1[ i ];
int  posj=dimBB+C2[ j ];
G12[ i ][ j ][ p]=Vp[ posi ][ p]*Vp[ posj ][ p];
}
}
}
}
//===================================================================//
void  mkP22()
{
int  C1[dimB ];
int  C2[dimB ];
for ( int  i =0; i <dimB ; i ++)  C1[ i ]= i ;
C2[0]=0;
for ( int  i =1; i <dimB ; i ++)  C2[ i ]=dimB−1+i ;
//for( int  i =0; i <dimB ; i ++)  printf("%d  %d\n", i , C2[ i ]);
for ( int  p =0; p<dimH ; p++)
{
P22[ p ]=0.0;
for ( int  i =0; i <dimB ; i ++)
{
for ( int  j =0; j <dimB ; j ++)
{
int  posi=dimBB+C2[ i ];
int  posj=dimBB+C2[ j ];
P22[ p]+=U[ i ]*Vp[ posi ][ p]*U[ j ]*Vp[ posj ][ p];
}
}
}
}
//===================================================================//
```

```cpp
void mkP12()
{
int  C1[dimB];
int  C2[dimB];
for(int  i=0;i<dimB;i++)  C1[i]=i;
C2[0]=0;
for(int  i=1;i<dimB;i++)  C2[i]=dimB-1+i;
//for(int  i=0;i<dimB;i++)  printf("%d %d\n",i,C2[i]);
for(int  p=0;p<dimH;p++)
{
P12[p]=0.0;
for(int  i=0;i<dimB;i++)
{
for(int  j=0;j<dimB;j++)
{
int  posi=C1[i];
int  posj=dimBB+C2[j];
P12[p]+=U[i]*Vp[posi][p]*U[j]*Vp[posj][p];
}
}
}
}
//=================================================================//
void mkG11_0000()
{
for(int  iw=0;iw<wpoints;iw++)
{
double  sr=0.0;
double  si=0.0;
for(int  p=0;p<dimH;p++)
{
double  tampr=wtab[iw]-Ep[p];
double  tampi=eta;
double  denom=tampr*tampr+tampi*tampi;
sr+=G11[0][0][p]*tampr/denom;
si-=G11[0][0][p]*tampi/denom;
}
G11_0000[0][iw]=sr;
G11_0000[1][iw]=si;
}
}
//=================================================================//
void mkG12_0000()
{
for(int  iw=0;iw<wpoints;iw++)
{
double  sr=0.0;
double  si=0.0;
for(int  p=0;p<dimH;p++)
{
double  tampr=wtab[iw]-Ep[p];
```

```cpp
double tampi=eta;
double denom=tampr*tampr+tampi*tampi;
sr+=G12[0][0][p]*tampr/denom;
si-=G12[0][0][p]*tampi/denom;
}
G12_0000[0][iw]=sr;
G12_0000[1][iw]=si;
}
}
//=============================================================//
void mkP12_0000()
{
for(int iw=0;iw<wpoints;iw++)
{
double sr=0.0;
double si=0.0;
for(int p=0;p<dimH;p++)
{
double tampr=wtab[iw]-Ep[p];
double tampi=eta;
double denom=tampr*tampr+tampi*tampi;
sr+=P12[p]*tampr/denom;
si-=P12[p]*tampi/denom;
}
P12_0000[0][iw]=sr;
P12_0000[1][iw]=si;
}
}
//=============================================================//
void mkP22_0000()
{
for(int iw=0;iw<wpoints;iw++)
{
double sr=0.0;
double si=0.0;
for(int p=0;p<dimH;p++)
{
double tampr=wtab[iw]-Ep[p];
double tampi=eta;
double denom=tampr*tampr+tampi*tampi;
sr+=P22[p]*tampr/denom;
si-=P22[p]*tampi/denom;
}
P22_0000[0][iw]=sr;
P22_0000[1][iw]=si;
}
}
//=============================================================//
void saveG11_0000()
{
FILE *fp;
fp=fopen("G11e.dat","w");
```

```cpp
for(int iw=0;iw<wpoints;iw++)
  fprintf(fp,"%18.18f_%18.18f_%18.18f\n",wtab[iw],G11_0000[0][iw],
                                         G11_0000[1][iw]);

fclose(fp);
}
//================================================================//
void saveG12_0000()
{
FILE *fp;
fp=fopen("G12e.dat","w");
for(int iw=0;iw<wpoints;iw++)
  fprintf(fp,"%18.18f_%18.18f_%18.18f\n",wtab[iw],G12_0000[0][iw],
                                         G12_0000[1][iw]);

fclose(fp);
}
//================================================================//
void saveP12_0000()
{
FILE *fp;
fp=fopen("P12e.dat","w");
for(int iw=0;iw<wpoints;iw++)
  fprintf(fp,"%18.18f_%18.18f_%18.18f\n",wtab[iw],P12_0000[0][iw],
                                         P12_0000[1][iw]);

fclose(fp);
}
//================================================================//
void saveP22_0000()
{
FILE *fp;
fp=fopen("P22e.dat","w");
for(int iw=0;iw<wpoints;iw++)
  fprintf(fp,"%18.18f_%18.18f_%18.18f\n",wtab[iw],P22_0000[0][iw],
                                         P22_0000[1][iw]);

fclose(fp);
}
//================================================================//

main()
{
mkNB();
mkV1();
mkV2();
//seeH1();
//seeH2();
mkD1();
mkD2();
//seeH1();
//seeH2();
mkHD();
mkHT();
//seeH();
diagH();
```

```cpp
  mkG11();
  mkG12();
  for(int iw=0;iw<wpoints;iw++)  wtab[iw]=wbeg+iw*wpas;
  mkG11_0000();
  mkG12_0000();
  saveG11_0000();
  saveG12_0000();
  mkU();
  mkP12();
  mkP22();
  mkP12_0000();
  mkP22_0000();
  saveP12_0000();
  saveP22_0000();
  }
//===================================================================//
```

twoG.cc

```cpp
//===================================================================//
//twoG.cc
//===================================================================//
#include <cmath>
#include <cstdio>
#include <cstdlib>
//===================================================================//
const double e0=0.0;
const double w0=1.0;
const double t0=1.0;
const double g0=1.0;
//===================================================================//
const int dimB=10;
//===================================================================//
const double wbeg=-10.0;
const double wend=10.0;
const int wpoints=2000;
const double wpas=(wend-wbeg)/wpoints;
const double eta=0.1;
//===================================================================//
double wtab[wpoints];
double W0[2][wpoints],S0[2][wpoints],W1[2][wpoints];
double G0[2][wpoints],G[2][wpoints],G12[2][wpoints];
//===================================================================//
int Inear(double a)
{
double a1=floor(a);
double a2=ceil(a);
if(fabs(a1-a)<fabs(a2-a)) return int(a1);
else return int(a2);
}
//===================================================================//
void mkW0()
```

```cpp
{
for(int  iw=0;iw<wpoints ;iw++)
{
double  tampr=wtab[iw]−e0 ;
double  tampi=eta ;
double  denom=tampr*tampr+tampi*tampi ;
W0[0][iw]=t0*t0*tampr/denom ;
W0[1][iw]=−t0*t0*tampi/denom ;
}// endfor  iw
}
//=================================================================//
void  mkW1()
{
for(int  iw=0;iw<wpoints ;iw++)
{
double  tampr=wtab[iw]−e0−S0[0][iw] ;
double  tampi=eta−S0[1][iw] ;
double  denom=tampr*tampr+tampi*tampi ;
W1[0][iw]=t0*t0*tampr/denom ;
W1[1][iw]=−t0*t0*tampi/denom ;
}// endfor  iw
}
//=================================================================//
void  saveG11()
{
FILE *fp ;
fp=fopen("G11.dat","w");
for(int  iw=0;iw<wpoints ;iw++)
{
fprintf(fp ,"%18.18f_%18.18f_%18.18f\n",wtab[iw],G[0][iw],G[1][iw]);
}
fclose(fp );
}
//=================================================================//
void  saveG12()
{
FILE *fp ;
fp=fopen("G12.dat","w");
for(int  iw=0;iw<wpoints ;iw++)
{
fprintf(fp ,"%18.18f_%18.18f_%18.18f\n",wtab[iw],G12[0][iw],
                                             G12[1][iw]);

}
fclose(fp );
}
//=================================================================//
void  mkS0()
{
for(int  iw=0;iw<wpoints ;iw++)
{
double  sr=0.0;
double  si=0.0;
```

```cpp
for ( int  n=dimB ; n >0;n−−)
{
int  i2=Inear ( iw∗1.0−n∗w0/ wpas );
double  tampr=wtab [ iw]−e0−n∗w0−s r ;
double  tampi=eta−si ;
if (( i2 >=0)&&(i2 <wpoints )){ tampr −=W0[ 0 ][ i2 ] ; tampi −=W0[ 1 ][ i2 ] ; }
double  denom=tampr∗tampr+tampi∗tampi ;
s r=n∗g0∗g0∗w0∗w0∗tampr / denom ;
s i=−n∗g0∗g0∗w0∗w0∗tampi / denom ;
}// end  n
S0 [ 0 ][ iw]= s r ;
S0 [ 1 ][ iw]= s i ;
}// end  iw
}
//=================================================================//
void  mkG0 ( )
{
for ( int  iw =0;iw <wpoints ; iw ++)
{
double  tampr=wtab [ iw]−e0−S0 [ 0 ][ iw ] ;
double  tampi=eta −S0 [ 1 ][ iw ] ;
double  denom=tampr∗tampr+tampi∗tampi ;
G0 [ 0 ][ iw ]= tampr / denom ;
G0 [ 1 ][ iw]=− tampi / denom ;
}
}
//=================================================================//
void  mkG ( )
{
for ( int  iw =0;iw <wpoints ; iw ++)
{
double  tampr=wtab [ iw]−e0−S0 [ 0 ][ iw]−W1[ 0 ][ iw ] ;
double  tampi=eta −S0 [ 1 ][ iw]−W1[ 1 ][ iw ] ;
double  denom=tampr∗tampr+tampi∗tampi ;
G[ 0 ][ iw ]= tampr / denom ;
G[ 1 ][ iw]=− tampi / denom ;
}
}
//=================================================================//
void  mkG12 ( )
{
for ( int  iw =0;iw <wpoints ; iw ++)
{
double  tampr=G0 [ 0 ][ iw ]∗G[ 0 ][ iw]−G0 [ 1 ][ iw ]∗G[ 1 ][ iw ] ;
double  tampi=G0 [ 0 ][ iw ]∗G[ 1 ][ iw ]+G0 [ 1 ][ iw ]∗G[ 0 ][ iw ] ;
G12 [ 0 ][ iw]=− t0 ∗tampr ;
G12 [ 1 ][ iw]=− t0 ∗tampi ;
}
}
//=================================================================//
main ( )
{
```

```cpp
for(int iw=0;iw<wpoints;iw++)  wtab[iw]=wbeg+iw*wpas;
mkW0();
mkS0();
mkW1();
mkG0();
mkG();
mkG12();
saveG11();
saveG12();
}
//===============================================================//
```

twoP.cc

```cpp
//===============================================================//
// twoP . cc
//===============================================================//
#include <cmath>
#include <cstdio>
#include <cstdlib>
#include "ludcmp.h"
#include "lubksb.h"
#include "mprove.h"
//===============================================================//
const double e0=0.0;
const double w0=1.0;
const double t0=1.0;
const double g0=1.0;
const double pi=acos(-1.0);
//===============================================================//
const int dimB=10;
//===============================================================//
const double wbeg=-10.0;
const double wend=10.0;
const int wpoints=2000;
const double wpas=(wend-wbeg)/wpoints;
const double eta=0.1;
//===============================================================//
int iz;
double wtab[wpoints];
double W0[2][wpoints],W1[2][wpoints],S0[2][wpoints];
double U[dimB];
double K11r[dimB][dimB],K11i[dimB][dimB];
double G11r[dimB][dimB],G11i[dimB][dimB];
double K22r[dimB][dimB],K22i[dimB][dimB];
double G22r[dimB][dimB],G22i[dimB][dimB];
double P12[2][wpoints];
//===============================================================//
int Inear(double a)
{
double a1=floor(a);
double a2=ceil(a);
```

```cpp
if ( fabs ( a1 −a)< fabs ( a2 −a )) return int ( a1 );
else return int ( a2 );
}
//===================================================================//
void mkU()
{
double tamp=exp ( −0.5 ∗ g0 ∗ g0 );
U[ 0 ]= tamp ;
for ( int n=1;n<dimB ; n++)
{
tamp=tamp ∗ g0 / sqrt ( 1.0 ∗ n );
U[ n ]= tamp ;
}
}
//===================================================================//
void mkW0()
{
for ( int iw =0; iw<wpoints ; iw ++)
{
double tampr=wtab [ iw ]− e0 ;
double tampi=eta ;
double denom=tampr ∗ tampr+tampi ∗ tampi ;
W0[ 0 ] [ iw ]= t0 ∗ t0 ∗ tampr / denom ;
W0[ 1 ] [ iw ]=− t0 ∗ t0 ∗ tampi / denom ;
} // endfor iw
}
//===================================================================//
void mkW1()
{
for ( int iw =0; iw<wpoints ; iw ++)
{
double tampr=wtab [ iw ]− e0 −S0 [ 0 ] [ iw ];
double tampi=eta −S0 [ 1 ] [ iw ];
double denom=tampr ∗ tampr+tampi ∗ tampi ;
W1[ 0 ] [ iw ]= t0 ∗ t0 ∗ tampr / denom ;
W1[ 1 ] [ iw ]=− t0 ∗ t0 ∗ tampi / denom ;
} // endfor iw
}
//===================================================================//
void mkS0()
{
for ( int iw =0; iw<wpoints ; iw ++)
{
double sr =0.0;
double si =0.0;
for ( int n=dimB ; n>0; n−−)
{
int i2 =Inear ( iw ∗ 1.0 −n ∗ w0/ wpas );
double tampr=wtab [ iw ]− e0 −n ∗ w0−sr ;
double tampi=eta −si ;
if (( i2 >=0)&&(i2 <wpoints )){ tampr −=W0[ 0 ] [ i2 ]; tampi −=W0[ 1 ] [ i2 ]; }
double denom=tampr ∗ tampr+tampi ∗ tampi ;
```

```cpp
sr=n*g0*g0*w0*w0*tampr/denom;
si=-n*g0*g0*w0*w0*tampi/denom;
}//end n
S0[0][iw]=sr;
S0[1][iw]=si;
}//end iw
}
//==================================================================//
void mkK11()
{
for(int i=0;i<dimB;i++) for(int j=0;j<dimB;j++) K11r[i][j]=0.0;
for(int i=0;i<dimB;i++) for(int j=0;j<dimB;j++) K11i[i][j]=0.0;
for(int n=0;n<dimB;n++) K11r[n][n]=wtab[iz]-e0-n*w0;
for(int n=0;n<dimB;n++) K11i[n][n]=eta;
for(int i=1;i<dimB;i++) K11r[i-1][i]=K11r[i][i-1]=g0*w0*sqrt(1.0*i);
K11r[0][0]-=W1[0][iz];
K11i[0][0]-=W1[1][iz];
for(int n=1;n<dimB;n++)
{
int i2=Inear(iz*1.0-n*w0/wpas);
if(i2>=0){K11r[n][n]-=W0[0][i2];K11i[n][n]-=W0[1][i2];}
}
}
//==================================================================//
void mkK22()
{
for(int i=0;i<dimB;i++) for(int j=0;j<dimB;j++) K22r[i][j]=0.0;
for(int i=0;i<dimB;i++) for(int j=0;j<dimB;j++) K22i[i][j]=0.0;
for(int n=0;n<dimB;n++) K22r[n][n]=wtab[iz]-e0-n*w0;
for(int n=0;n<dimB;n++) K22i[n][n]=eta;
for(int i=1;i<dimB;i++) K22r[i-1][i]=K22r[i][i-1]=g0*w0*sqrt(1.0*i);
for(int n=1;n<dimB;n++)
{
int i2=Inear(iz*1.0-n*w0/wpas);
if(i2>=0){K22r[n][n]-=W0[0][i2];K22i[n][n]-=W0[1][i2];}
}
}
//==================================================================//
void mkG11()
{
double **g_mat;g_mat=new double*[2*dimB];
for(int i=0;i<2*dimB;i++) g_mat[i]=new double[2*dimB];
double **glu_mat;glu_mat=new double*[2*dimB];
for(int i=0;i<2*dimB;i++) glu_mat[i]=new double[2*dimB];
double *x_vec=new double[2*dimB];
double *b_vec=new double[2*dimB];
int *perm_vec=new int[2*dimB];
double dperm;
double err;

for(int i=0;i<dimB;i++)
{
```

```cpp
for(int  j=0;j<dimB;j++)
{
g_mat[i][j]=K11r[i][j];
g_mat[i][j+dimB]=-K11i[i][j];
g_mat[i+dimB][j]=K11i[i][j];
g_mat[i+dimB][j+dimB]=K11r[i][j];
}
}

for(int  i=0;i<2*dimB;i++)  for(int  j=0;j<2*dimB;j++)
glu_mat[i][j]=g_mat[i][j];

ludcmp(glu_mat,2*dimB,perm_vec,dperm);

for(int  n=0;n<dimB;n++)
{
for(int  i=0;i<2*dimB;i++)  x_vec[i]=0.0;x_vec[n]=1.0;
lubksb(glu_mat,2*dimB,perm_vec,x_vec);
for(int  i=0;i<2*dimB;i++)  b_vec[i]=0.0;b_vec[n]=1.0;
do
{
mprove(g_mat,glu_mat,2*dimB,perm_vec,b_vec,x_vec,err);
}while(err>10e-12);
for(int  i=0;i<dimB;i++)  G11r[i][n]=x_vec[i];
for(int  i=0;i<dimB;i++)  G11i[i][n]=x_vec[i+dimB];
}

for(int  i=0;i<2*dimB;i++)  delete[]  g_mat[i];delete[]  g_mat;
for(int  i=0;i<2*dimB;i++)  delete[]  glu_mat[i];delete[]  glu_mat;
delete[]  x_vec;
delete[]  b_vec;
delete[]  perm_vec;
}
//================================================================//
void  mkG22()
{
double  **g_mat;g_mat=new  double*[2*dimB];
for(int  i=0;i<2*dimB;i++)  g_mat[i]=new  double[2*dimB];
double  **glu_mat;glu_mat=new  double*[2*dimB];
for(int  i=0;i<2*dimB;i++)  glu_mat[i]=new  double[2*dimB];
double  *x_vec=new  double[2*dimB];
double  *b_vec=new  double[2*dimB];
int  *perm_vec=new  int[2*dimB];
double  dperm;
double  err;

for(int  i=0;i<dimB;i++)
{
for(int  j=0;j<dimB;j++)
{
g_mat[i][j]=K22r[i][j];
g_mat[i][j+dimB]=-K22i[i][j];
```

```cpp
g_mat[i+dimB][j]=K22i[i][j];
g_mat[i+dimB][j+dimB]=K22r[i][j];
}
}

for(int i=0;i<2*dimB;i++) for(int j=0;j<2*dimB;j++)
glu_mat[i][j]=g_mat[i][j];

ludcmp(glu_mat,2*dimB,perm_vec,dperm);

for(int n=0;n<dimB;n++)
{
for(int i=0;i<2*dimB;i++) x_vec[i]=0.0;x_vec[n]=1.0;
lubksb(glu_mat,2*dimB,perm_vec,x_vec);
for(int i=0;i<2*dimB;i++) b_vec[i]=0.0;b_vec[n]=1.0;
do
{
mprove(g_mat,glu_mat,2*dimB,perm_vec,b_vec,x_vec,err);
}while(err>10e-12);
for(int i=0;i<dimB;i++) G22r[i][n]=x_vec[i];
for(int i=0;i<dimB;i++) G22i[i][n]=x_vec[i+dimB];
}

for(int i=0;i<2*dimB;i++) delete[] g_mat[i];delete[] g_mat;
for(int i=0;i<2*dimB;i++) delete[] glu_mat[i];delete[] glu_mat;
delete[] x_vec;
delete[] b_vec;
delete[] perm_vec;
}
//================================================================//
void mkP12()
{
double tamp1r=0.0;
double tamp1i=0.0;
for(int i=0;i<dimB;i++)
{
tamp1r+=U[i]*G11r[i][0];
tamp1i+=U[i]*G11i[i][0];
}
double tamp2r=0.0;
double tamp2i=0.0;
for(int i=0;i<dimB;i++)
{
tamp2r+=U[i]*G22r[0][i];
tamp2i+=U[i]*G22i[0][i];
}
double tampr=tamp1r*tamp2r-tamp1i*tamp2i;
double tampi=tamp1r*tamp2i+tamp1i*tamp2r;
P12[0][iz]=-t0*tampr;
P12[1][iz]=-t0*tampi;
}
//================================================================//
```

```cpp
void saveP12()
{
FILE *fp;
fp=fopen("P12.dat","w");
for(int iw=0;iw<wpoints;iw++)
fprintf(fp,"%18.18f_%18.18f_%18.18f\n",wtab[iw],P12[0][iw],
                                                P12[1][iw]);
fclose(fp);
}
//================================================================//
main()
{
for(int iw=0;iw<wpoints;iw++)  wtab[iw]=wbeg+iw*wpas;
mkW0();
mkS0();
mkW1();
mkU();
for(iz=0;iz<wpoints;iz++)
{
mkK11();
mkG11();
mkK22();
mkG22();
mkP12();
}
saveP12();
}
//================================================================//
```

dmft2.cc

```cpp
//================================================================//
// dmft2.cc
//================================================================//
#include <cmath>
#include <cstdio>
#include <cstdlib>
#include "ludcmp.h"
#include "lubksb.h"
#include "mprove.h"
//================================================================//
const double e0=0.0;
const double w0=1.0;
const double t0=1.0;
const double g0=0.9;
const double pi=acos(-1.0);
//================================================================//
const int dimB=20;
//================================================================//
const double wbeg=-10.0;
const double wend=10.0;
const int wpoints=2000;
```

```cpp
const double wpas=(wend-wbeg)/wpoints;
const double eta=0.1;
//==========================================================//
int iz;
double Erreur;
double wtab[wpoints];
double W[2][wpoints],S[2][wpoints],Sold[2][wpoints];
double G11a[2][wpoints],G12a[2][wpoints];
double G11e[2][wpoints],G12e[2][wpoints];
double P11e[2][wpoints],P12e[2][wpoints];
double K11r[dimB][dimB],K11i[dimB][dimB];
double G11r[dimB][dimB],G11i[dimB][dimB];
double K22r[dimB][dimB],K22i[dimB][dimB];
double G22r[dimB][dimB],G22i[dimB][dimB];
double U[dimB];
//==========================================================//
int Inear(double a)
{
double a1=floor(a);
double a2=ceil(a);
if(fabs(a1-a)<fabs(a2-a)) return int(a1);
else return int(a2);
}
//==========================================================//
void initS()
{
for(int iw=0;iw<wpoints;iw++)
{
Sold[0][iw]=0.0;
Sold[1][iw]=0.0;
}
}
//==========================================================//
void mkW()
{
for(int iw=0;iw<wpoints;iw++)
{
double tampr=wtab[iw]-e0-Sold[0][iw];
double tampi=eta-Sold[1][iw];
double denom=tampr*tampr+tampi*tampi;
W[0][iw]=t0*t0*tampr/denom;
W[1][iw]=-t0*t0*tampi/denom;
}//endfor iw
}
//==========================================================//
void mkS()
{
for(int iw=0;iw<wpoints;iw++)
{
double sr=0.0;
double si=0.0;
for(int n=dimB;n>0;n--)
```

```cpp
{
int  i2=Inear(iw*1.0-n*w0/wpas);
double  tampr=wtab[iw]-e0-n*w0-sr;
double  tampi=eta-si;
if(((i2>=0)&&(i2<wpoints)){ tampr-=W[0][i2]; tampi-=W[1][i2];}
double  denom=tampr*tampr+tampi*tampi;
sr=n*g0*g0*w0*w0*tampr/denom;
si=-n*g0*g0*w0*w0*tampi/denom;
}//end  n
S[0][iw]=sr;
S[1][iw]=si;
}//end  iw
}
//=================================================================//
void  mkError()
{
Erreur=0.0;
for(int  iw=0;iw<wpoints;iw++)
{
Erreur+=fabs(S[0][iw]-Sold[0][iw])+fabs(S[1][iw]-Sold[1][iw]);
Sold[0][iw]=0.5*Sold[0][iw]+0.5*S[0][iw];
Sold[1][iw]=0.5*Sold[1][iw]+0.5*S[1][iw];
}
}
//=================================================================//
void  mkG11a()
{
for(int  iw=0;iw<wpoints;iw++)
{
double  tampr=wtab[iw]-e0-S[0][iw]-W[0][iw];
double  tampi=eta-S[1][iw]-W[1][iw];
double  denom=tampr*tampr+tampi*tampi;
G11a[0][iw]=tampr/denom;
G11a[1][iw]=-tampi/denom;
}
}
//=================================================================//
void  mkG12a()
{
for(int  iw=0;iw<wpoints;iw++)
{
double  tampr=W[0][iw]*G11a[0][iw]-W[1][iw]*G11a[1][iw];
double  tampi=W[0][iw]*G11a[1][iw]+W[1][iw]*G11a[0][iw];
G12a[0][iw]=-t0*tampr;
G12a[1][iw]=-t0*tampi;
}
}
//=================================================================//
void  saveG11a()
{
FILE  *fp;
fp=fopen("G11a.dat","w");
```

```cpp
for(int iw=0;iw<wpoints;iw++)
fprintf(fp,"%18.18f_%18.18f_%18.18f\n",wtab[iw],G11a[0][iw],
                                        G11a[1][iw]);
fclose(fp);
}
//================================================================//
void saveG12a()
{
FILE *fp;
fp=fopen("G12a.dat","w");
for(int iw=0;iw<wpoints;iw++)
fprintf(fp,"%18.18f_%18.18f_%18.18f\n",wtab[iw],G12a[0][iw],
                                        G12a[1][iw]);
fclose(fp);
}
//================================================================//
void mkK11()
{
for(int i=0;i<dimB;i++) for(int j=0;j<dimB;j++) K11r[i][j]=0.0;
for(int i=0;i<dimB;i++) for(int j=0;j<dimB;j++) K11i[i][j]=0.0;
for(int n=0;n<dimB;n++) K11r[n][n]=wtab[iz]-e0-n*w0;
for(int n=0;n<dimB;n++) K11i[n][n]=eta;
for(int i=1;i<dimB;i++) K11r[i-1][i]=K11r[i][i-1]=g0*w0*sqrt(1.0*i);
for(int n=0;n<dimB;n++)
{
int i2=Inear(iz*1.0-n*w0/wpas);
if(i2>=0){K11r[n][n]-=W[0][i2]; K11i[n][n]-=W[1][i2];}
}
}
//================================================================//
void mkK22()
{
for(int i=0;i<dimB;i++) for(int j=0;j<dimB;j++) K22r[i][j]=0.0;
for(int i=0;i<dimB;i++) for(int j=0;j<dimB;j++) K22i[i][j]=0.0;
for(int n=0;n<dimB;n++) K22r[n][n]=wtab[iz]-e0-n*w0;
for(int n=0;n<dimB;n++) K22i[n][n]=eta;
for(int i=1;i<dimB;i++) K22r[i-1][i]=K22r[i][i-1]=g0*w0*sqrt(1.0*i);
for(int n=1;n<dimB;n++)
{
int i2=Inear(iz*1.0-n*w0/wpas);
if(i2>=0){K22r[n][n]-=W[0][i2]; K22i[n][n]-=W[1][i2];}
}
}
//================================================================//
void mkG11()
{
double **g_mat;g_mat=new double*[2*dimB];
for(int i=0;i<2*dimB;i++) g_mat[i]=new double[2*dimB];
double **glu_mat;glu_mat=new double*[2*dimB];
for(int i=0;i<2*dimB;i++) glu_mat[i]=new double[2*dimB];
double *x_vec=new double[2*dimB];
double *b_vec=new double[2*dimB];
```

```cpp
int *perm_vec=new int[2*dimB];
double dperm;
double err;

for(int i=0;i<dimB;i++)
{
for(int j=0;j<dimB;j++)
{
g_mat[i][j]=K11r[i][j];
g_mat[i][j+dimB]=-K11i[i][j];
g_mat[i+dimB][j]=K11i[i][j];
g_mat[i+dimB][j+dimB]=K11r[i][j];
}
}

for(int i=0;i<2*dimB;i++) for(int j=0;j<2*dimB;j++)
glu_mat[i][j]=g_mat[i][j];

ludcmp(glu_mat,2*dimB,perm_vec,dperm);

for(int n=0;n<dimB;n++)
{
for(int i=0;i<2*dimB;i++) x_vec[i]=0.0;x_vec[n]=1.0;
lubksb(glu_mat,2*dimB,perm_vec,x_vec);
for(int i=0;i<2*dimB;i++) b_vec[i]=0.0;b_vec[n]=1.0;
do
{
mprove(g_mat,glu_mat,2*dimB,perm_vec,b_vec,x_vec,err);
}while(err>10e-12);
for(int i=0;i<dimB;i++) G11r[i][n]=x_vec[i];
for(int i=0;i<dimB;i++) G11i[i][n]=x_vec[i+dimB];
}

for(int i=0;i<2*dimB;i++) delete[] g_mat[i];delete[] g_mat;
for(int i=0;i<2*dimB;i++) delete[] glu_mat[i];delete[] glu_mat;
delete[] x_vec;
delete[] b_vec;
delete[] perm_vec;
}
//================================================================//
void mkG22()
{
double **g_mat;g_mat=new double*[2*dimB];
for(int i=0;i<2*dimB;i++) g_mat[i]=new double[2*dimB];
double **glu_mat;glu_mat=new double*[2*dimB];
for(int i=0;i<2*dimB;i++) glu_mat[i]=new double[2*dimB];
double *x_vec=new double[2*dimB];
double *b_vec=new double[2*dimB];
int *perm_vec=new int[2*dimB];
double dperm;
double err;
```

```cpp
for(int  i=0;i<dimB;i++)
{
for(int  j=0;j<dimB;j++)
{
g_mat[i][j]=K22r[i][j];
g_mat[i][j+dimB]=-K22i[i][j];
g_mat[i+dimB][j]=K22i[i][j];
g_mat[i+dimB][j+dimB]=K22r[i][j];
}
}

for(int  i=0;i<2*dimB;i++)  for(int  j=0;j<2*dimB;j++)
glu_mat[i][j]=g_mat[i][j];

ludcmp(glu_mat,2*dimB,perm_vec,dperm);

for(int  n=0;n<dimB;n++)
{
for(int  i=0;i<2*dimB;i++)  x_vec[i]=0.0;x_vec[n]=1.0;
lubksb(glu_mat,2*dimB,perm_vec,x_vec);
for(int  i=0;i<2*dimB;i++)  b_vec[i]=0.0;b_vec[n]=1.0;
do
{
mprove(g_mat,glu_mat,2*dimB,perm_vec,b_vec,x_vec,err);
}while(err>10e-12);
for(int  i=0;i<dimB;i++)  G22r[i][n]=x_vec[i];
for(int  i=0;i<dimB;i++)  G22i[i][n]=x_vec[i+dimB];
}

for(int  i=0;i<2*dimB;i++)  delete []  g_mat[i];delete []  g_mat;
for(int  i=0;i<2*dimB;i++)  delete []  glu_mat[i];delete []  glu_mat;
delete []  x_vec;
delete []  b_vec;
delete []  perm_vec;
}
//=================================================================//
void mkU()
{
double  tamp=exp(-0.5*g0*g0);
U[0]=tamp;
for(int  n=1;n<dimB;n++)
{
tamp=tamp*g0/sqrt(1.0*n);
U[n]=tamp;
}
}
//=================================================================//
void mkG11e()
{
G11e[0][iz]=G11r[0][0];
G11e[1][iz]=G11i[0][0];
}
```

```cpp
//================================================================//
void mkG12e ( )
{
double  tampr=G11r[0][0]*G22r[0][0] − G11i [0][0]*G22i [0][0];
double  tampi=G11r[0][0]*G22i[0][0]+G11i[0][0]*G22r[0][0];
G12e [0][ iz]=− t0*tampr ;
G12e [1][ iz]=− t0*tampi ;
}
//================================================================//
void  mkP11e ( )
{
P11e [0][ iz ]=0.0;
P11e [1][ iz ]=0.0;
for ( int  m=0;m<dimB ;m++)
{
for ( int  n =0;n<dimB ;n++)
{
P11e [0][ iz ]+=U[m]*U[ n]*G11r [m][ n ];
P11e [1][ iz ]+=U[m]*U[ n]*G11i [m][ n ];
}
}
}
//================================================================//
void  mkP12e ( )
{
double  tamp1r =0.0;
double  tamp1i =0.0;
for ( int  i =0; i <dimB ; i ++)
{
tamp1r+=U[ i ]*G11r [ i ][0];
tamp1i+=U[ i ]*G11i [ i ][0];
}
double  tamp2r =0.0;
double  tamp2i =0.0;
for ( int  i =0; i <dimB ; i ++)
{
tamp2r+=U[ i ]*G22r [0][ i ];
tamp2i+=U[ i ]*G22i [0][ i ];
}
double  tampr=tamp1r*tamp2r−tamp1i*tamp2i ;
double  tampi=tamp1r*tamp2i+tamp1i*tamp2r ;
P12e [0][ iz]=− t0*tampr ;
P12e [1][ iz]=− t0*tampi ;
}
//================================================================//
void  saveG11e ( )
{
FILE  *fp ;
fp=fopen ( ”G11e . dat ” , ”w” );
for ( int  iw =0;iw<wpoints ; iw ++)
fprintf ( fp , ”%18.18 f_%18.18 f_%18.18 f\n” , wtab [ iw ] , G11e [0][ iw ] ,
                                              G11e [1][ iw ]);
```

```cpp
fclose(fp);
}
//=================================================================//
void saveG12e()
{
FILE *fp;
fp=fopen("G12e.dat","w");
for(int iw=0;iw<wpoints;iw++)
fprintf(fp,"%18.18f_%18.18f_%18.18f\n",wtab[iw],G12e[0][iw],
                                            G12e[1][iw]);
fclose(fp);
}
//=================================================================//
void saveP11e()
{
FILE *fp;
fp=fopen("P11e.dat","w");
for(int iw=0;iw<wpoints;iw++)
fprintf(fp,"%18.18f_%18.18f_%18.18f\n",wtab[iw],P11e[0][iw],
                                            P11e[1][iw]);
fclose(fp);
}
//=================================================================//
void saveP12e()
{
FILE *fp;
fp=fopen("P12e.dat","w");
for(int iw=0;iw<wpoints;iw++)
fprintf(fp,"%18.18f_%18.18f_%18.18f\n",wtab[iw],P12e[0][iw],
                                            P12e[1][iw]);
fclose(fp);
}
//=================================================================//
void savePk()
{
FILE *fp;
fp=fopen("P0.dat","w");
for(int iw=0;iw<wpoints;iw++)
{
double tampr=P11e[0][iw]+P12e[0][iw];
double tampi=P11e[1][iw]+P12e[1][iw];
fprintf(fp,"%18.18f_%18.18f_%18.18f\n",wtab[iw],tampr,tampi);
}
fclose(fp);
fp=fopen("Ppi.dat","w");
for(int iw=0;iw<wpoints;iw++)
{
double tampr=P11e[0][iw]-P12e[0][iw];
double tampi=P11e[1][iw]-P12e[1][iw];
fprintf(fp,"%18.18f_%18.18f_%18.18f\n",wtab[iw],tampr,tampi);
}
fclose(fp);
```

```cpp
}
//=================================================================//
main ()
{
for ( int  iw =0;iw<wpoints ; iw ++)   wtab [ iw ]=wbeg+iw∗wpas ;
initS ();
do
{
mkW2 ();
mkS ();
mkError ();
printf ("Erreur=%18.18f\n" , Erreur );
} while ( Erreur >10e −6);
mkG11a ();
mkG12a ();
saveG11a ();
saveG12a ();
mkU ();
for ( iz =0;iz<wpoints ; iz ++)
{
mkK11 ();
mkG11 ();
mkK22 ();
mkG22 ();
mkG11e ();
mkG12e ();
mkP11e ();
mkP12e ();
}
saveG11e ();
saveG12e ();
saveP11e ();
saveP12e ();
savePk ();
}
//=================================================================//
```

exact3.cc

```cpp
//=================================================================//
// exact3 . cc
//=================================================================//
// Holstein  3  sites  OPEN CHAIN
// restricted  basis  no  crossing
//=================================================================//
#include  <cmath>
#include  <cstdio>
#include  <cstdlib>
#include  "jacobi .h"
#include  "eigsrt .h"
//=================================================================//
const  double  e0 =0.0;
```

```cpp
const double w0=1.0;
const double t0 =1.0;
const double g0 =1.0;
const int dimB =10;
const int dimBB=3*(dimB −1)+1;
const int dimH=3*dimBB;
//================================================================//
const double wbeg= −10.0;
const double wend =10.0;
const int wpoints =2000;
const double wpas =(wend−wbeg)/ wpoints ;
const double eta =0.1;
//================================================================//
int B1[dimB],B2[dimB],B3[dimB];
int C1[dimB],C2[dimB],C3[dimB];
int NB[dimBB];
double U[dimB];
double H1[dimBB][dimBB],H2[dimBB][dimBB],H3[dimBB][dimBB];
double H[dimH][dimH],Ep[dimH],Vp[dimH][dimH];
double G11[dimB][dimB][dimH],G12[dimB][dimB][dimH];
double G13[dimB][dimB][dimH],G22[dimB][dimB][dimH];
double P11[dimH],P12[dimH],P13[dimH],P22[dimH];
double wtab[wpoints];
double G11_00[2][wpoints],G12_00[2][wpoints],G13_00[2][wpoints];
double G22_00[2][wpoints];
double P11_00[2][wpoints],P12_00[2][wpoints],P13_00[2][wpoints];
double P22_00[2][wpoints];
//================================================================//
void mkU()
{
double tamp=exp(−0.5*g0*g0);
U[0]=tamp;
for(int n=1;n<dimB;n++)
{
tamp=tamp*g0/sqrt(1.0*n);
U[n]=tamp;
}
//for(int i=0;i<dimB;i++) printf("i=%d U=%lf\n",i,U[i]);
}
//================================================================//
void mkB()
{
for(int i=0;i<dimB;i++) B1[i]=B2[i]=B3[i]=0;
for(int i=0;i<dimB;i++) C1[i]=C2[i]=C3[i]=0;
for(int i=0;i<dimB;i++) B1[i]=i;
B2[0]=0;for(int i=1;i<dimB;i++) B2[i]=i+dimB−1;
B3[0]=0;for(int i=1;i<dimB;i++) B3[i]=i+2*(dimB−1);
// printf("B:\n");
//for(int i=0;i<dimB;i++) printf("%d %d %d\n",B1[i],B2[i],B3[i]);
for(int i=0;i<dimB;i++) C1[i]=B1[i];
for(int i=0;i<dimB;i++) C2[i]=B2[i]+dimBB;
for(int i=0;i<dimB;i++) C3[i]=B3[i]+2*dimBB;
```

```cpp
//printf("C:\n");
//for(int i=0;i<dimB;i++) printf("%d %d %d\n",C1[i],C2[i],C3[i]);
}
//================================================================//
void mkNB()
{
for(int i=0;i<dimB;i++)
{
int pos1=B1[i];NB[pos1]=i;
int pos2=B2[i];NB[pos2]=i;
int pos3=B3[i];NB[pos3]=i;
}
}
//================================================================//
void mkV1()
{
for(int i=0;i<dimBB;i++) for(int j=0;j<dimBB;j++) H1[i][j]=0.0;
for(int i=1;i<dimB;i++) H1[i-1][i]=H1[i][i-1]=sqrt(1.0*i);
}
//================================================================//
void mkV2()
{
for(int i=0;i<dimBB;i++) for(int j=0;j<dimBB;j++) H2[i][j]=0.0;
for(int i=1;i<dimB;i++)
{
int posi=B2[i];
int posj=B2[i-1];
H2[posi][posj]=H2[posj][posi]=sqrt(1.0*i);
}
}
//================================================================//
void mkV3()
{
for(int i=0;i<dimBB;i++) for(int j=0;j<dimBB;j++) H3[i][j]=0.0;
for(int i=1;i<dimB;i++)
{
int posi=B3[i];
int posj=B3[i-1];
H3[posi][posj]=H3[posj][posi]=sqrt(1.0*i);
}
}
//================================================================//
void mkD1()
{
for(int i=0;i<dimBB;i++) for(int j=0;j<dimBB;j++)
H1[i][j]=-g0*w0*H1[i][j];
for(int i=0;i<dimBB;i++) H1[i][i]+=e0+w0*NB[i];
}
//================================================================//
void mkD2()
{
for(int i=0;i<dimBB;i++) for(int j=0;j<dimBB;j++)
```

```cpp
H2[ i ][ j ]=−g0*w0*H2[ i ][ j ];
for(int  i =0;i<dimBB ; i ++)  H2[ i ][ i ]+=e0+w0*NB[ i ];
}
//==================================================================//
void  mkD3 ()
{
for(int  i =0;i<dimBB ; i ++)  for(int  j =0;j<dimBB ; j ++)
H3[ i ][ j ]=−g0*w0*H3[ i ][ j ];
for(int  i =0;i<dimBB ; i ++)  H3[ i ][ i ]+=e0+w0*NB[ i ];
}
//==================================================================//
void  seeH1 ()
{
printf ("H1\n");
for(int  i =0;i<dimBB ; i ++)
{
for(int  j =0;j<dimBB ; j ++)  printf ("%3.2f_",H1[ i ][ j ]);
printf ("\n");
}
}
//==================================================================//
void  seeH2 ()
{
printf ("H2\n");
for(int  i =0;i<dimBB ; i ++)
{
for(int  j =0;j<dimBB ; j ++)  printf ("%3.2f_",H2[ i ][ j ]);
printf ("\n");
}
}
//==================================================================//
void  seeH3 ()
{
printf ("H3\n");
for(int  i =0;i<dimBB ; i ++)
{
for(int  j =0;j<dimBB ; j ++)  printf ("%3.2f_",H3[ i ][ j ]);
printf ("\n");
}
}
//==================================================================//
void  mkHD ()
{
for(int  i =0;i<dimH ; i ++)  for(int  j =0;j<dimH ; j ++)  H[ i ][ j ]=0.0;
for(int  i =0;i<dimBB ; i ++)
{
for(int  j =0;j<dimBB ; j ++)
{
H[ i ][ j ]=H1[ i ][ j ];
H[ i +dimBB ][ j +dimBB ]=H2[ i ][ j ];
H[ i +2*dimBB ][ j +2*dimBB ]=H3[ i ][ j ];
}
```

```cpp
}
}
//===================================================================//
void mkHT()
{
for(int i=0;i<dimBB;i++)
{
H[dimBB+i][i]=H[i][dimBB+i]=-t0;
H[2*dimBB+i][dimBB+i]=H[dimBB+i][2*dimBB+i]=-t0;
}
}
//===================================================================//
void seeH()
{
printf("H=\n");
for(int i=0;i<dimH;i++)
{
for(int j=0;j<dimH;j++) printf("%3.2f ",H[i][j]);
printf("\n");
}
}
//===================================================================//
void diagH()
{
double **h_mat;h_mat=new double *[dimH];
for(int i=0;i<dimH;i++) h_mat[i]=new double[dimH];
double **v_mat;v_mat=new double *[dimH];
for(int i=0;i<dimH;i++) v_mat[i]=new double[dimH];
double *d_vec=new double[dimH];

for(int i=0;i<dimH;i++) for(int j=0;j<dimH;j++) h_mat[i][j]=H[i][j];
jacobi(h_mat,dimH,d_vec,v_mat,10e-24);
eigsrt(d_vec,v_mat,dimH);
for(int i=0;i<dimH;i++) Ep[i]=d_vec[i];
for(int i=0;i<dimH;i++) for(int j=0;j<dimH;j++)
                        Vp[i][j]=v_mat[i][j];

for(int i=0;i<dimH;i++) delete [] h_mat[i];delete [] h_mat;
for(int i=0;i<dimH;i++) delete [] v_mat[i];delete [] v_mat;
delete [] d_vec;
}
//===================================================================//
void mkG11()
{
for(int i=0;i<dimB;i++)
{
for(int j=0;j<dimB;j++)
{
for(int p=0;p<dimH;p++)
{
int posi=C1[i];
int posj=C1[j];
```

```cpp
G11[ i ][ j ][ p]=Vp[ posi ][ p]*Vp[ posj ][ p ];
}
}
}
}
//================================================================//
void mkG12()
{
for(int  i =0;i<dimB ; i ++)
{
for(int  j =0;j<dimB ; j ++)
{
for(int  p=0;p<dimH;p++)
{
int  posi=C1[ i ];
int  posj=C2[ j ];
G12[ i ][ j ][ p]=Vp[ posi ][ p]*Vp[ posj ][ p ];
}
}
}
}
//================================================================//
void mkG13()
{
for(int  i =0;i<dimB ; i ++)
{
for(int  j =0;j<dimB ; j ++)
{
for(int  p=0;p<dimH;p++)
{
int  posi=C1[ i ];
int  posj=C3[ j ];
G13[ i ][ j ][ p]=Vp[ posi ][ p]*Vp[ posj ][ p ];
}
}
}
}
//================================================================//
void mkG22()
{
for(int  i =0;i<dimB ; i ++)
{
for(int  j =0;j<dimB ; j ++)
{
for(int  p=0;p<dimH;p++)
{
int  posi=C2[ i ];
int  posj=C2[ j ];
G22[ i ][ j ][ p]=Vp[ posi ][ p]*Vp[ posj ][ p ];
}
}
}
```

```cpp
}
//==================================================================//
void mkP11()
{
for(int p=0;p<dimH;p++)
{
P11[p]=0.0;
for(int i=0;i<dimB;i++)
{
for(int j=0;j<dimB;j++)
{
int posi=C1[i];
int posj=C1[j];
P11[p]+=U[i]*Vp[posi][p]*U[j]*Vp[posj][p];
}
}
}
}
//==================================================================//
void mkP12()
{
for(int p=0;p<dimH;p++)
{
P12[p]=0.0;
for(int i=0;i<dimB;i++)
{
for(int j=0;j<dimB;j++)
{
int posi=C1[i];
int posj=C2[j];
P12[p]+=U[i]*Vp[posi][p]*U[j]*Vp[posj][p];
}
}
}
}
//==================================================================//
void mkP13()
{
for(int p=0;p<dimH;p++)
{
P13[p]=0.0;
for(int i=0;i<dimB;i++)
{
for(int j=0;j<dimB;j++)
{
int posi=C1[i];
int posj=C3[j];
P13[p]+=U[i]*Vp[posi][p]*U[j]*Vp[posj][p];
}
}
}
}
```

```cpp
//===============================================================//
void mkP22()
{
for(int p=0;p<dimH;p++)
{
P22[p]=0.0;
for(int i=0;i<dimB;i++)
{
for(int j=0;j<dimB;j++)
{
int posi=C2[i];
int posj=C2[j];
P22[p]+=U[i]*Vp[posi][p]*U[j]*Vp[posj][p];
}
}
}
}
//===============================================================//
void mkG11_00()
{
for(int iw=0;iw<wpoints;iw++)
{
double sr=0.0;
double si=0.0;
for(int p=0;p<dimH;p++)
{
double tampr=wtab[iw]-Ep[p];
double tampi=eta;
double denom=tampr*tampr+tampi*tampi;
sr+=G11[0][0][p]*tampr/denom;
si-=G11[0][0][p]*tampi/denom;
}
G11_00[0][iw]=sr;
G11_00[1][iw]=si;
}
}
//===============================================================//
void mkG12_00()
{
for(int iw=0;iw<wpoints;iw++)
{
double sr=0.0;
double si=0.0;
for(int p=0;p<dimH;p++)
{
double tampr=wtab[iw]-Ep[p];
double tampi=eta;
double denom=tampr*tampr+tampi*tampi;
sr+=G12[0][0][p]*tampr/denom;
si-=G12[0][0][p]*tampi/denom;
}
G12_00[0][iw]=sr;
```

```cpp
G12_00 [ 1 ] [ iw ]= si ;
}
}
//=================================================================//
void  mkG13_00 ( )
{
for ( int  iw =0; iw<wpoints ; iw ++)
{
double  sr =0.0;
double  si =0.0;
for ( int  p=0; p<dimH ; p++)
{
double  tampr=wtab [ iw ]−Ep[ p ] ;
double  tampi=eta ;
double  denom=tampr*tampr+tampi*tampi ;
sr+=G13 [ 0 ] [ 0 ] [ p ]* tampr / denom ;
si −=G13 [ 0 ] [ 0 ] [ p ]* tampi / denom ;
}
G13_00 [ 0 ] [ iw ]= sr ;
G13_00 [ 1 ] [ iw ]= si ;
}
}
//=================================================================//
void  mkG22_00 ( )
{
for ( int  iw =0; iw<wpoints ; iw ++)
{
double  sr =0.0;
double  si =0.0;
for ( int  p=0; p<dimH ; p++)
{
double  tampr=wtab [ iw ]−Ep[ p ] ;
double  tampi=eta ;
double  denom=tampr*tampr+tampi*tampi ;
sr+=G22 [ 0 ] [ 0 ] [ p ]* tampr / denom ;
si −=G22 [ 0 ] [ 0 ] [ p ]* tampi / denom ;
}
G22_00 [ 0 ] [ iw ]= sr ;
G22_00 [ 1 ] [ iw ]= si ;
}
}
//=================================================================//
void  mkP11_00 ( )
{
for ( int  iw =0; iw<wpoints ; iw ++)
{
double  sr =0.0;
double  si =0.0;
for ( int  p=0; p<dimH ; p++)
{
double  tampr=wtab [ iw ]−Ep[ p ] ;
double  tampi=eta ;
```

```cpp
double denom=tampr*tampr+tampi*tampi;
sr+=P11[p]*tampr/denom;
si-=P11[p]*tampi/denom;
}
P11_00[0][iw]=sr;
P11_00[1][iw]=si;
}
}
//===================================================================//
void mkP12_00()
{
for(int iw=0;iw<wpoints;iw++)
{
double sr=0.0;
double si=0.0;
for(int p=0;p<dimH;p++)
{
double tampr=wtab[iw]-Ep[p];
double tampi=eta;
double denom=tampr*tampr+tampi*tampi;
sr+=P12[p]*tampr/denom;
si-=P12[p]*tampi/denom;
}
P12_00[0][iw]=sr;
P12_00[1][iw]=si;
}
}
//===================================================================//
void mkP13_00()
{
for(int iw=0;iw<wpoints;iw++)
{
double sr=0.0;
double si=0.0;
for(int p=0;p<dimH;p++)
{
double tampr=wtab[iw]-Ep[p];
double tampi=eta;
double denom=tampr*tampr+tampi*tampi;
sr+=P13[p]*tampr/denom;
si-=P13[p]*tampi/denom;
}
P13_00[0][iw]=sr;
P13_00[1][iw]=si;
}
}
//===================================================================//
void mkP22_00()
{
for(int iw=0;iw<wpoints;iw++)
{
double sr=0.0;
```

```cpp
double  si =0.0;
for(int  p=0;p<dimH;p++)
{
double  tampr=wtab[iw]-Ep[p];
double  tampi=eta;
double  denom=tampr*tampr+tampi*tampi;
sr+=P22[p]*tampr/denom;
si-=P22[p]*tampi/denom;
}
P22_00[0][iw]=sr;
P22_00[1][iw]=si;
}
}
//================================================================//
void  saveG11_00()
{
FILE *fp;
fp=fopen("G11e.dat","w");
for(int  iw=0;iw<wpoints;iw++)
fprintf(fp,"%18.18f_%18.18f_%18.18f\n",wtab[iw],G11_00[0][iw],
                                       G11_00[1][iw]);
fclose(fp);
}
//================================================================//
void  saveG12_00()
{
FILE *fp;
fp=fopen("G12e.dat","w");
for(int  iw=0;iw<wpoints;iw++)
fprintf(fp,"%18.18f_%18.18f_%18.18f\n",wtab[iw],G12_00[0][iw],
                                       G12_00[1][iw]);
fclose(fp);
}
//================================================================//
void  saveG13_00()
{
FILE *fp;
fp=fopen("G13e.dat","w");
for(int  iw=0;iw<wpoints;iw++)
fprintf(fp,"%18.18f_%18.18f_%18.18f\n",wtab[iw],G13_00[0][iw],
                                       G13_00[1][iw]);
fclose(fp);
}
//================================================================//
void  saveG22_00()
{
FILE *fp;
fp=fopen("G22e.dat","w");
for(int  iw=0;iw<wpoints;iw++)
fprintf(fp,"%18.18f_%18.18f_%18.18f\n",wtab[iw],G22_00[0][iw],
                                       G22_00[1][iw]);
fclose(fp);
```

```cpp
}
//==================================================================//
void saveP11_00()
{
FILE *fp;
fp=fopen("P11e.dat","w");
for(int iw=0;iw<wpoints;iw++)
fprintf(fp,"%18.18f_%18.18f_%18.18f\n",wtab[iw],P11_00[0][iw],
                                       P11_00[1][iw]);

fclose(fp);
}
//==================================================================//
void saveP12_00()
{
FILE *fp;
fp=fopen("P12e.dat","w");
for(int iw=0;iw<wpoints;iw++)
fprintf(fp,"%18.18f_%18.18f_%18.18f\n",wtab[iw],P12_00[0][iw],
                                       P12_00[1][iw]);

fclose(fp);
}
//==================================================================//
void saveP13_00()
{
FILE *fp;
fp=fopen("P13e.dat","w");
for(int iw=0;iw<wpoints;iw++)
fprintf(fp,"%18.18f_%18.18f_%18.18f\n",wtab[iw],P13_00[0][iw],
                                       P13_00[1][iw]);

fclose(fp);
}
//==================================================================//
void saveP22_00()
{
FILE *fp;
fp=fopen("P22e.dat","w");
for(int iw=0;iw<wpoints;iw++)
fprintf(fp,"%18.18f_%18.18f_%18.18f\n",wtab[iw],P22_00[0][iw],
                                       P22_00[1][iw]);

fclose(fp);
}
//==================================================================//

main()
{
printf("dimB=%d_dimBB=%d_dimH=%d\n",dimB,dimBB,dimH);
mkB();
mkNB();
mkV1();mkV2();mkV3();
//seeH1();seeH2();seeH3();
mkD1();mkD2();mkD3();
//seeH1();seeH2();seeH3();
```

```
mkHD ( ) ;
mkHT ( ) ;
//seeH ( ) ;
diagH ( ) ;
mkG11 ( ) ;
mkG12 ( ) ;
mkG13 ( ) ;
mkG22 ( ) ;
for ( int iw=0;iw<wpoints ;iw++)  wtab [iw]=wbeg+iw∗wpas ;
mkG11_00 ( ) ;
mkG12_00 ( ) ;
mkG13_00 ( ) ;
mkG22_00 ( ) ;
saveG11_00 ( ) ;
saveG12_00 ( ) ;
saveG13_00 ( ) ;
saveG22_00 ( ) ;
mkU ( ) ;
mkP11 ( ) ;
mkP12 ( ) ;
mkP13 ( ) ;
mkP22 ( ) ;
mkP11_00 ( ) ;
mkP12_00 ( ) ;
mkP13_00 ( ) ;
mkP22_00 ( ) ;
saveP11_00 ( ) ;
saveP12_00 ( ) ;
saveP13_00 ( ) ;
saveP22_00 ( ) ;
}
```

threeG.cc

```
//=================================================================//
// threeG . cc
//=================================================================//
#include <cmath>
#include <cstdio>
#include <cstdlib>
//=================================================================//
const double e0 =0.0;
const double w0=1.0;
const double t0=1.0;
const double g0=1.0;
//=================================================================//
const int dimB =10;
//=================================================================//
const double wbeg=−10.0;
const double wend=10.0;
const int wpoints =2000;
const double wpas =(wend−wbeg )/ wpoints ;
```

```cpp
const double eta=0.1;
//==============================================================//
double wtab[wpoints];
double W10[2][wpoints],S10[2][wpoints],W11[2][wpoints];
double W20[2][wpoints],S20[2][wpoints],W21[2][wpoints];
double G11[2][wpoints],G22[2][wpoints];
double W221[2][wpoints],G221[2][wpoints],G12[2][wpoints];
double G3312[2][wpoints],G13[2][wpoints];
//==============================================================//
int Inear(double a)
{
double a1=floor(a);
double a2=ceil(a);
if(fabs(a1-a)<fabs(a2-a)) return int(a1);
else return int(a2);
}
//==============================================================//
void mkW10()
{
for(int iw=0;iw<wpoints;iw++)
{
double tr=wtab[iw]-e0;
double ti=eta;
double denom=tr*tr+ti*ti;
tr=tr/denom;
ti=-ti/denom;
tr=wtab[iw]-e0-t0*t0*tr;
ti=eta-t0*t0*ti;
denom=tr*tr+ti*ti;
tr=tr/denom;
ti=-ti/denom;
W10[0][iw]=t0*t0*tr;
W10[1][iw]=t0*t0*ti;
}
}
//==============================================================//
void mkW11()
{
for(int iw=0;iw<wpoints;iw++)
{
double tr=wtab[iw]-e0-S10[0][iw];
double ti=eta-S10[1][iw];
double denom=tr*tr+ti*ti;
tr=tr/denom;
ti=-ti/denom;
tr=wtab[iw]-e0-S20[0][iw]-t0*t0*tr;
ti=eta-S20[1][iw]-t0*t0*ti;
denom=tr*tr+ti*ti;
tr=tr/denom;
ti=-ti/denom;
W11[0][iw]=t0*t0*tr;
W11[1][iw]=t0*t0*ti;
```

```cpp
}
}
//================================================================//
void mkW20()
{
for(int iw=0;iw<wpoints;iw++)
{
double tr=wtab[iw]-e0;
double ti=eta;
double denom=tr*tr+ti*ti;
tr=tr/denom;
ti=-ti/denom;
W20[0][iw]=2.0*t0*t0*tr;
W20[1][iw]=2.0*t0*t0*ti;
}
}
//================================================================//
void mkW21()
{
for(int iw=0;iw<wpoints;iw++)
{
double tr=wtab[iw]-e0-S10[0][iw];
double ti=eta-S10[1][iw];
double denom=tr*tr+ti*ti;
tr=tr/denom;
ti=-ti/denom;
W21[0][iw]=2.0*t0*t0*tr;
W21[1][iw]=2.0*t0*t0*ti;
}
}
//================================================================//
void mkW221()
{
for(int iw=0;iw<wpoints;iw++)
{
double tr=wtab[iw]-e0-S10[0][iw];
double ti=eta-S10[1][iw];
double denom=tr*tr+ti*ti;
tr=tr/denom;
ti=-ti/denom;
W221[0][iw]=t0*t0*tr;
W221[1][iw]=t0*t0*ti;
}
}
//================================================================//
void mkS10()
{
for(int iw=0;iw<wpoints;iw++)
{
double sr=0.0;
double si=0.0;
for(int n=dimB;n>0;n--)
```

```cpp
{
int i2=Inear(iw*1.0-n*w0/wpas);
double tampr=wtab[iw]-e0-n*w0-sr;
double tampi=eta-si;
if((i2>=0)&&(i2<wpoints)){tampr-=W10[0][i2];tampi-=W10[1][i2];}
double denom=tampr*tampr+tampi*tampi;
sr=n*g0*g0*w0*w0*tampr/denom;
si=-n*g0*g0*w0*w0*tampi/denom;
}//end n
S10[0][iw]=sr;
S10[1][iw]=si;
}//end iw
}
//===============================================================//
void mkS20()
{
for(int iw=0;iw<wpoints;iw++)
{
double sr=0.0;
double si=0.0;
for(int n=dimB;n>0;n--)
{
int i2=Inear(iw*1.0-n*w0/wpas);
double tampr=wtab[iw]-e0-n*w0-sr;
double tampi=eta-si;
if((i2>=0)&&(i2<wpoints)){tampr-=W20[0][i2];tampi-=W20[1][i2];}
double denom=tampr*tampr+tampi*tampi;
sr=n*g0*g0*w0*w0*tampr/denom;
si=-n*g0*g0*w0*w0*tampi/denom;
}//end n
S20[0][iw]=sr;
S20[1][iw]=si;
}//end iw
}
//===============================================================//
void mkG11()
{
for(int iw=0;iw<wpoints;iw++)
{
double tampr=wtab[iw]-e0-S10[0][iw]-W11[0][iw];
double tampi=eta-S10[1][iw]-W11[1][iw];
double denom=tampr*tampr+tampi*tampi;
tampr=tampr/denom;
tampi=-tampi/denom;
G11[0][iw]=tampr;
G11[1][iw]=tampi;
}
}
//===============================================================//
void mkG22()
{
for(int iw=0;iw<wpoints;iw++)
```

```cpp
{
double  tampr=wtab[iw]-e0-S20[0][iw]-W21[0][iw];
double  tampi=eta-S20[1][iw]-W21[1][iw];
double  denom=tampr*tampr+tampi*tampi;
tampr=tampr/denom;
tampi=-tampi/denom;
G22[0][iw]=tampr;
G22[1][iw]=tampi;
}
}
//======================================================================//
void  mkG221()
{
for(int  iw=0;iw<wpoints;iw++)
{
double  tampr=wtab[iw]-e0-S20[0][iw]-W221[0][iw];
double  tampi=eta-S20[1][iw]-W221[1][iw];
double  denom=tampr*tampr+tampi*tampi;
tampr=tampr/denom;
tampi=-tampi/denom;
G221[0][iw]=tampr;
G221[1][iw]=tampi;
}
}
//======================================================================//

void  mkG12()
{
for(int  iw=0;iw<wpoints;iw++)
{
double  tampr=G11[0][iw]*G221[0][iw]-G11[1][iw]*G221[1][iw];
double  tampi=G11[0][iw]*G221[1][iw]+G11[1][iw]*G221[0][iw];
G12[0][iw]=-t0*tampr;
G12[1][iw]=-t0*tampi;
}
}
//======================================================================//
void  mkG3312()
{
for(int  iw=0;iw<wpoints;iw++)
{
double  tampr=wtab[iw]-e0-S10[0][iw];
double  tampi=eta-S10[1][iw];
double  denom=tampr*tampr+tampi*tampi;
tampr=tampr/denom;
tampi=-tampi/denom;
G3312[0][iw]=tampr;
G3312[1][iw]=tampi;
}
}
//======================================================================//
void  mkG13()
```

```cpp
{
for(int iw=0;iw<wpoints;iw++)
{
double tampr=G12[0][iw]*G3312[0][iw]-G12[1][iw]*G3312[1][iw];
double tampi=G12[0][iw]*G3312[1][iw]+G12[1][iw]*G3312[0][iw];
G13[0][iw]=-t0*tampr;
G13[1][iw]=-t0*tampi;
}
}
//================================================================//

void saveG11()
{
FILE *fp;
fp=fopen("G11a.dat","w");
for(int iw=0;iw<wpoints;iw++)
fprintf(fp,"%18.18f_%18.18f_%18.18f\n",wtab[iw],G11[0][iw],
                                            G11[1][iw]);

fclose(fp);
}
//================================================================//
void saveG22()
{
FILE *fp;
fp=fopen("G22a.dat","w");
for(int iw=0;iw<wpoints;iw++)
fprintf(fp,"%18.18f_%18.18f_%18.18f\n",wtab[iw],G22[0][iw],
                                            G22[1][iw]);

fclose(fp);
}
//================================================================//
void saveG12()
{
FILE *fp;
fp=fopen("G12a.dat","w");
for(int iw=0;iw<wpoints;iw++)
fprintf(fp,"%18.18f_%18.18f_%18.18f\n",wtab[iw],G12[0][iw],
                                            G12[1][iw]);

fclose(fp);
}
//================================================================//
void saveG13()
{
FILE *fp;
fp=fopen("G13a.dat","w");
for(int iw=0;iw<wpoints;iw++)
fprintf(fp,"%18.18f_%18.18f_%18.18f\n",wtab[iw],G13[0][iw],
                                            G13[1][iw]);

fclose(fp);
}
//================================================================//
main()
```

```cpp
{
for(int iw=0;iw<wpoints;iw++)  wtab[iw]=wbeg+iw*wpas;
mkW10();
mkW20();
mkS10();
mkS20();
mkW11();
mkW21();
mkG11();
saveG11();
mkG22();
saveG22();
mkW221();
mkG221();
mkG12();
saveG12();
mkG3312();
mkG13();
saveG13();
}
//================================================================//
```

threeP.cc

```cpp
//================================================================//
//threeP.cc
//================================================================//
#include <cmath>
#include <cstdio>
#include <cstdlib>
#include "ludcmp.h"
#include "lubksb.h"
#include "mprove.h"
//================================================================//
const double e0=0.0;
const double w0=1.0;
const double t0=1.0;
const double g0=1.0;
const double pi=acos(-1.0);
//================================================================//
const int dimB=10;
//================================================================//
const double wbeg=-10.0;
const double wend=10.0;
const int wpoints=2000;
const double wpas=(wend-wbeg)/wpoints;
const double eta=0.1;
//================================================================//
int iz;
double wtab[wpoints];
double W10[2][wpoints],S10[2][wpoints],W11[2][wpoints];
double W20[2][wpoints],S20[2][wpoints],W21[2][wpoints];
```

```cpp
double W221[2][wpoints];
double U[dimB];
double K11r[dimB][dimB],K11i[dimB][dimB];
double G11r[dimB][dimB],G11i[dimB][dimB];
double P11[2][wpoints];
double K22r[dimB][dimB],K22i[dimB][dimB];
double G22r[dimB][dimB],G22i[dimB][dimB];
double P22[2][wpoints];
double K221r[dimB][dimB],K221i[dimB][dimB];
double G221r[dimB][dimB],G221i[dimB][dimB];
double P12[2][wpoints];
double K3312r[dimB][dimB],K3312i[dimB][dimB];
double G3312r[dimB][dimB],G3312i[dimB][dimB];
double P13[2][wpoints];
//==================================================================//
void mkU()
{
double tamp=exp(-0.5*g0*g0);
U[0]=tamp;
for(int n=1;n<dimB;n++)
{
tamp=tamp*g0/sqrt(1.0*n);
U[n]=tamp;
}
}
//==================================================================//
int Inear(double a)
{
double a1=floor(a);
double a2=ceil(a);
if(fabs(a1-a)<fabs(a2-a)) return int(a1);
else return int(a2);
}
//==================================================================//
void mkW10()
{
for(int iw=0;iw<wpoints;iw++)
{
double tr=wtab[iw]-e0;
double ti=eta;
double denom=tr*tr+ti*ti;
tr=tr/denom;
ti=-ti/denom;
tr=wtab[iw]-e0-t0*t0*tr;
ti=eta-t0*t0*ti;
denom=tr*tr+ti*ti;
tr=tr/denom;
ti=-ti/denom;
W10[0][iw]=t0*t0*tr;
W10[1][iw]=t0*t0*ti;
}
}
```

```cpp
//================================================================//
void mkW11()
{
for(int iw=0;iw<wpoints;iw++)
{
double tr=wtab[iw]-e0-S10[0][iw];
double ti=eta-S10[1][iw];
double denom=tr*tr+ti*ti;
tr=tr/denom;
ti=-ti/denom;
tr=wtab[iw]-e0-S20[0][iw]-t0*t0*tr;
ti=eta-S20[1][iw]-t0*t0*ti;
denom=tr*tr+ti*ti;
tr=tr/denom;
ti=-ti/denom;
W11[0][iw]=t0*t0*tr;
W11[1][iw]=t0*t0*ti;
}
}
//================================================================//
void mkW20()
{
for(int iw=0;iw<wpoints;iw++)
{
double tr=wtab[iw]-e0;
double ti=eta;
double denom=tr*tr+ti*ti;
tr=tr/denom;
ti=-ti/denom;
W20[0][iw]=2.0*t0*t0*tr;
W20[1][iw]=2.0*t0*t0*ti;
}
}
//================================================================//
void mkW21()
{
for(int iw=0;iw<wpoints;iw++)
{
double tr=wtab[iw]-e0-S10[0][iw];
double ti=eta-S10[1][iw];
double denom=tr*tr+ti*ti;
tr=tr/denom;
ti=-ti/denom;
W21[0][iw]=2.0*t0*t0*tr;
W21[1][iw]=2.0*t0*t0*ti;
}
}
//================================================================//
void mkW221()
{
for(int iw=0;iw<wpoints;iw++)
{
```

```cpp
double  tr=wtab[iw]−e0−S10[0][iw];
double  ti=eta−S10[1][iw];
double  denom=tr*tr+ti*ti;
tr=tr/denom;
ti=−ti/denom;
W221[0][iw]=t0*t0*tr;
W221[1][iw]=t0*t0*ti;
}
}
//===============================================================//
void  mkS10()
{
for(int  iw=0;iw<wpoints;iw++)
{
double  sr=0.0;
double  si=0.0;
for(int  n=dimB;n>0;n−−)
{
int  i2=Inear(iw*1.0−n*w0/wpas);
double  tampr=wtab[iw]−e0−n*w0−sr;
double  tampi=eta−si;
if((i2>=0)&&(i2<wpoints)){tampr−=W10[0][i2];tampi−=W10[1][i2];}
double  denom=tampr*tampr+tampi*tampi;
sr=n*g0*g0*w0*w0*tampr/denom;
si=−n*g0*g0*w0*w0*tampi/denom;
}//end  n
S10[0][iw]=sr;
S10[1][iw]=si;
}//end  iw
}
//===============================================================//
void  mkS20()
{
for(int  iw=0;iw<wpoints;iw++)
{
double  sr=0.0;
double  si=0.0;
for(int  n=dimB;n>0;n−−)
{
int  i2=Inear(iw*1.0−n*w0/wpas);
double  tampr=wtab[iw]−e0−n*w0−sr;
double  tampi=eta−si;
if((i2>=0)&&(i2<wpoints)){tampr−=W20[0][i2];tampi−=W20[1][i2];}
double  denom=tampr*tampr+tampi*tampi;
sr=n*g0*g0*w0*w0*tampr/denom;
si=−n*g0*g0*w0*w0*tampi/denom;
}//end  n
S20[0][iw]=sr;
S20[1][iw]=si;
}//end  iw
}
//===============================================================//
```

```cpp
void mkK11()
{
for(int i=0;i<dimB;i++)  for(int j=0;j<dimB;j++)  K11r[i][j]=0.0;
for(int i=0;i<dimB;i++)  for(int j=0;j<dimB;j++)  K11i[i][j]=0.0;
for(int n=0;n<dimB;n++)  K11r[n][n]=wtab[iz]-e0-n*w0;
for(int n=0;n<dimB;n++)  K11i[n][n]=eta;
for(int i=1;i<dimB;i++)  K11r[i-1][i]=K11r[i][i-1]=g0*w0*sqrt(1.0*i);
K11r[0][0]-=W11[0][iz];
K11i[0][0]-=W11[1][iz];
for(int n=1;n<dimB;n++)
{
int i2=Inear(iz*1.0-n*w0/wpas);
if(i2>=0){K11r[n][n]-=W10[0][i2]; K11i[n][n]-=W10[1][i2];}
}
}
//===============================================================//
void mkK22()
{
for(int i=0;i<dimB;i++)  for(int j=0;j<dimB;j++)  K22r[i][j]=0.0;
for(int i=0;i<dimB;i++)  for(int j=0;j<dimB;j++)  K22i[i][j]=0.0;
for(int n=0;n<dimB;n++)  K22r[n][n]=wtab[iz]-e0-n*w0;
for(int n=0;n<dimB;n++)  K22i[n][n]=eta;
for(int i=1;i<dimB;i++)  K22r[i-1][i]=K22r[i][i-1]=g0*w0*sqrt(1.0*i);
K22r[0][0]-=W21[0][iz];
K22i[0][0]-=W21[1][iz];
for(int n=1;n<dimB;n++)
{
int i2=Inear(iz*1.0-n*w0/wpas);
if(i2>=0){K22r[n][n]-=W20[0][i2]; K22i[n][n]-=W20[1][i2];}
}
}
//===============================================================//
void mkK221()
{
for(int i=0;i<dimB;i++)  for(int j=0;j<dimB;j++)  K221r[i][j]=0.0;
for(int i=0;i<dimB;i++)  for(int j=0;j<dimB;j++)  K221i[i][j]=0.0;
for(int n=0;n<dimB;n++)  K221r[n][n]=wtab[iz]-e0-n*w0;
for(int n=0;n<dimB;n++)  K221i[n][n]=eta;
for(int i=1;i<dimB;i++)
   K221r[i-1][i]=K221r[i][i-1]=g0*w0*sqrt(1.0*i);
K221r[0][0]-=W221[0][iz];
K221i[0][0]-=W221[1][iz];
for(int n=1;n<dimB;n++)
{
int i2=Inear(iz*1.0-n*w0/wpas);
if(i2>=0){K221r[n][n]-=W20[0][i2]; K221i[n][n]-=W20[1][i2];}
}
}
//===============================================================//
void mkK3312()
{
for(int i=0;i<dimB;i++)  for(int j=0;j<dimB;j++)  K3312r[i][j]=0.0;
```

```cpp
for(int  i=0;i<dimB;i++)  for(int  j=0;j<dimB;j++) K3312i[i][j]=0.0;
for(int  n=0;n<dimB;n++) K3312r[n][n]=wtab[iz]-e0-n*w0;
for(int  n=0;n<dimB;n++) K3312i[n][n]=eta;
for(int  i=1;i<dimB;i++)
    K3312r[i-1][i]=K3312r[i][i-1]=g0*w0*sqrt(1.0*i);
for(int  n=1;n<dimB;n++)
{
int i2=Inear(iz*1.0-n*w0/wpas);
if(i2>=0){K3312r[n][n]-=W10[0][i2];K3312i[n][n]-=W10[1][i2];}
}
}
//===================================================================//
void mkG11()
{
double **g_mat;g_mat=new double*[2*dimB];
for(int  i=0;i<2*dimB;i++) g_mat[i]=new double[2*dimB];
double **glu_mat;glu_mat=new double*[2*dimB];
for(int  i=0;i<2*dimB;i++) glu_mat[i]=new double[2*dimB];
double *x_vec=new double[2*dimB];
double *b_vec=new double[2*dimB];
int *perm_vec=new int[2*dimB];
double dperm;
double err;

for(int  i=0;i<dimB;i++)
{
for(int  j=0;j<dimB;j++)
{
g_mat[i][j]=K11r[i][j];
g_mat[i][j+dimB]=-K11i[i][j];
g_mat[i+dimB][j]=K11i[i][j];
g_mat[i+dimB][j+dimB]=K11r[i][j];
}
}

for(int  i=0;i<2*dimB;i++)  for(int  j=0;j<2*dimB;j++)
glu_mat[i][j]=g_mat[i][j];

ludcmp(glu_mat,2*dimB,perm_vec,dperm);

for(int  n=0;n<dimB;n++)
{
for(int  i=0;i<2*dimB;i++) x_vec[i]=0.0;x_vec[n]=1.0;
lubksb(glu_mat,2*dimB,perm_vec,x_vec);
for(int  i=0;i<2*dimB;i++) b_vec[i]=0.0;b_vec[n]=1.0;
do
{
mprove(g_mat,glu_mat,2*dimB,perm_vec,b_vec,x_vec,err);
}while(err>10e-12);
for(int  i=0;i<dimB;i++) G11r[i][n]=x_vec[i];
for(int  i=0;i<dimB;i++) G11i[i][n]=x_vec[i+dimB];
}
```

```cpp
for(int i=0;i<2*dimB;i++) delete [] g_mat[i];delete [] g_mat;
for(int i=0;i<2*dimB;i++) delete [] glu_mat[i];delete [] glu_mat;
delete [] x_vec;
delete [] b_vec;
delete [] perm_vec;
}
//================================================================//
void mkG22()
{
double **g_mat;g_mat=new double*[2*dimB];
for(int i=0;i<2*dimB;i++) g_mat[i]=new double[2*dimB];
double **glu_mat;glu_mat=new double*[2*dimB];
for(int i=0;i<2*dimB;i++) glu_mat[i]=new double[2*dimB];
double *x_vec=new double[2*dimB];
double *b_vec=new double[2*dimB];
int *perm_vec=new int[2*dimB];
double dperm;
double err;

for(int i=0;i<dimB;i++)
{
for(int j=0;j<dimB;j++)
{
g_mat[i][j]=K22r[i][j];
g_mat[i][j+dimB]=-K22i[i][j];
g_mat[i+dimB][j]=K22i[i][j];
g_mat[i+dimB][j+dimB]=K22r[i][j];
}
}

for(int i=0;i<2*dimB;i++) for(int j=0;j<2*dimB;j++)
glu_mat[i][j]=g_mat[i][j];

ludcmp(glu_mat,2*dimB,perm_vec,dperm);

for(int n=0;n<dimB;n++)
{
for(int i=0;i<2*dimB;i++) x_vec[i]=0.0;x_vec[n]=1.0;
lubksb(glu_mat,2*dimB,perm_vec,x_vec);
for(int i=0;i<2*dimB;i++) b_vec[i]=0.0;b_vec[n]=1.0;
do
{
mprove(g_mat,glu_mat,2*dimB,perm_vec,b_vec,x_vec,err);
}while(err>10e-12);
for(int i=0;i<dimB;i++) G22r[i][n]=x_vec[i];
for(int i=0;i<dimB;i++) G22i[i][n]=x_vec[i+dimB];
}

for(int i=0;i<2*dimB;i++) delete [] g_mat[i];delete [] g_mat;
for(int i=0;i<2*dimB;i++) delete [] glu_mat[i];delete [] glu_mat;
delete [] x_vec;
```

```cpp
delete [] b_vec;
delete [] perm_vec;
}
//===================================================================//
void mkG221()
{
double **g_mat;g_mat=new double *[2*dimB];
for(int i=0;i<2*dimB;i++) g_mat[i]=new double[2*dimB];
double **glu_mat;glu_mat=new double *[2*dimB];
for(int i=0;i<2*dimB;i++) glu_mat[i]=new double[2*dimB];
double *x_vec=new double[2*dimB];
double *b_vec=new double[2*dimB];
int *perm_vec=new int[2*dimB];
double dperm;
double err;

for(int i=0;i<dimB;i++)
{
for(int j=0;j<dimB;j++)
{
g_mat[i][j]=K221r[i][j];
g_mat[i][j+dimB]=-K221i[i][j];
g_mat[i+dimB][j]=K221i[i][j];
g_mat[i+dimB][j+dimB]=K221r[i][j];
}
}

for(int i=0;i<2*dimB;i++) for(int j=0;j<2*dimB;j++)
glu_mat[i][j]=g_mat[i][j];

ludcmp(glu_mat,2*dimB,perm_vec,dperm);

for(int n=0;n<dimB;n++)
{
for(int i=0;i<2*dimB;i++) x_vec[i]=0.0;x_vec[n]=1.0;
lubksb(glu_mat,2*dimB,perm_vec,x_vec);
for(int i=0;i<2*dimB;i++) b_vec[i]=0.0;b_vec[n]=1.0;
do
{
mprove(g_mat,glu_mat,2*dimB,perm_vec,b_vec,x_vec,err);
}while(err>10e-12);
for(int i=0;i<dimB;i++) G221r[i][n]=x_vec[i];
for(int i=0;i<dimB;i++) G221i[i][n]=x_vec[i+dimB];
}

for(int i=0;i<2*dimB;i++) delete [] g_mat[i];delete [] g_mat;
for(int i=0;i<2*dimB;i++) delete [] glu_mat[i];delete [] glu_mat;
delete [] x_vec;
delete [] b_vec;
delete [] perm_vec;
}
//===================================================================//
```

```cpp
void mkG3312()
{
double **g_mat;g_mat=new double*[2*dimB];
for(int i=0;i<2*dimB;i++) g_mat[i]=new double[2*dimB];
double **glu_mat;glu_mat=new double*[2*dimB];
for(int i=0;i<2*dimB;i++) glu_mat[i]=new double[2*dimB];
double *x_vec=new double[2*dimB];
double *b_vec=new double[2*dimB];
int *perm_vec=new int[2*dimB];
double dperm;
double err;

for(int i=0;i<dimB;i++)
{
for(int j=0;j<dimB;j++)
{
g_mat[i][j]=K3312r[i][j];
g_mat[i][j+dimB]=-K3312i[i][j];
g_mat[i+dimB][j]=K3312i[i][j];
g_mat[i+dimB][j+dimB]=K3312r[i][j];
}
}

for(int i=0;i<2*dimB;i++) for(int j=0;j<2*dimB;j++)
glu_mat[i][j]=g_mat[i][j];

ludcmp(glu_mat,2*dimB,perm_vec,dperm);

for(int n=0;n<dimB;n++)
{
for(int i=0;i<2*dimB;i++) x_vec[i]=0.0;x_vec[n]=1.0;
lubksb(glu_mat,2*dimB,perm_vec,x_vec);
for(int i=0;i<2*dimB;i++) b_vec[i]=0.0;b_vec[n]=1.0;
do
{
mprove(g_mat,glu_mat,2*dimB,perm_vec,b_vec,x_vec,err);
}while(err>10e-12);
for(int i=0;i<dimB;i++) G3312r[i][n]=x_vec[i];
for(int i=0;i<dimB;i++) G3312i[i][n]=x_vec[i+dimB];
}

for(int i=0;i<2*dimB;i++) delete [] g_mat[i];delete [] g_mat;
for(int i=0;i<2*dimB;i++) delete [] glu_mat[i];delete [] glu_mat;
delete [] x_vec;
delete [] b_vec;
delete [] perm_vec;
}
//==================================================================//
void mkP11()
{
double tampr=0.0;
double tampi=0.0;
```

```cpp
for(int  i=0;i<dimB;i++)
{
for(int  j=0;j<dimB;j++)
{
tampr+=U[i]*G11r[i][j]*U[j];
tampi+=U[i]*G11i[i][j]*U[j];
}
}
P11[0][iz]=tampr;
P11[1][iz]=tampi;
}
//=================================================================//
void  mkP22()
{
double  tampr=0.0;
double  tampi=0.0;
for(int  i=0;i<dimB;i++)
{
for(int  j=0;j<dimB;j++)
{
tampr+=U[i]*G22r[i][j]*U[j];
tampi+=U[i]*G22i[i][j]*U[j];
}
}
P22[0][iz]=tampr;
P22[1][iz]=tampi;
}
//=================================================================//
void  mkP12()
{
double  tamp1r=0.0;
double  tamp1i=0.0;
for(int  i=0;i<dimB;i++)
{
tamp1r+=U[i]*G11r[i][0];
tamp1i+=U[i]*G11i[i][0];
}
double  tamp2r=0.0;
double  tamp2i=0.0;
for(int  i=0;i<dimB;i++)
{
tamp2r+=G221r[0][i]*U[i];
tamp2i+=G221i[0][i]*U[i];
}
double  tampr=tamp1r*tamp2r-tamp1i*tamp2i;
double  tampi=tamp1r*tamp2i+tamp1i*tamp2r;
P12[0][iz]=-t0*tampr;
P12[1][iz]=-t0*tampi;
}
//=================================================================//
void  mkP13()
{
```

```cpp
double  tamp1r =0.0;
double  tamp1i =0.0;
for ( int   i =0; i <dimB ; i ++)
{
tamp1r+=U[ i ]*G11r[ i ][0];
tamp1i+=U[ i ]*G11i[ i ][0];
}
double  tamp2r=G221r [0][0];
double  tamp2i=G221i [0][0];
double  tamp3r =0.0;
double  tamp3i =0.0;
for ( int   i =0; i <dimB ; i ++)
{
tamp3r+=G3312r [0][ i ]*U[ i ];
tamp3i+=G3312i [0][ i ]*U[ i ];
}
double  tampr=tamp1r*tamp2r-tamp1i*tamp2i ;
double  tampi=tamp1r*tamp2i+tamp1i*tamp2r ;
double  tr=tampr*tamp3r-tampi*tamp3i ;
double  ti=tampr*tamp3i+tampi*tamp3r ;
P13 [0][ iz ]=t0 *t0 *tr ;
P13 [1][ iz ]=t0 *t0 *ti ;
}
//================================================================//
void  saveP11 ()
{
FILE *fp ;
fp=fopen ("P11a . dat" ,"w" );
for ( int  iw =0; iw <wpoints ; iw ++)
fprintf ( fp , "%18.18 f_%18.18 f_%18.18 f\n" , wtab [ iw ] , P11 [0][ iw ] ,
                                            P11 [1][ iw ]);
fclose ( fp );
}
//================================================================//
void  saveP22 ()
{
FILE *fp ;
fp=fopen ("P22a . dat" ,"w" );
for ( int  iw =0; iw <wpoints ; iw ++)
fprintf ( fp , "%18.18 f_%18.18 f_%18.18 f\n" , wtab [ iw ] , P22 [0][ iw ] ,
                                            P22 [1][ iw ]);
fclose ( fp );
}
//================================================================//
void  saveP12 ()
{
FILE *fp ;
fp=fopen ("P12a . dat" ,"w" );
for ( int  iw =0; iw <wpoints ; iw ++)
fprintf ( fp , "%18.18 f_%18.18 f_%18.18 f\n" , wtab [ iw ] , P12 [0][ iw ] ,
                                            P12 [1][ iw ]);
fclose ( fp );
```

```cpp
}
//================================================================//
void saveP13()
{
FILE *fp;
fp=fopen("P13a.dat","w");
for(int iw=0;iw<wpoints;iw++)
fprintf(fp,"%18.18f_%18.18f_%18.18f\n",wtab[iw],P13[0][iw],
                                        P13[1][iw]);

fclose(fp);
}
//================================================================//
main()
{
for(int iw=0;iw<wpoints;iw++) wtab[iw]=wbeg+iw*wpas;
mkW10();
mkW20();
mkS10();
mkS20();
mkW11();
mkW21();
mkW221();
mkU();
for(iz=0;iz<wpoints;iz++)
{
mkK11();
mkG11();
mkP11();
mkK22();
mkG22();
mkP22();
mkK221();
mkG221();
mkP12();
mkK3312();
mkG3312();
mkP13();
}
saveP11();
saveP22();
saveP12();
saveP13();
}
//================================================================//
```

dmft11.cc

```cpp
//================================================================//
//dmft11.cc
//Spectral functions for k=0,1,2,3,4,5,6,7,8,9,10
//================================================================//
#include <cmath>
```

```cpp
#include <cstdio>
#include <cstdlib>
#include "ludcmp.h"
#include "lubksb.h"
#include "mprove.h"
//=================================================================//
const double e0=0.0;
const double w0=1.0;
const double t0=1.0;
const double g0=1.0;
const double pi=acos(-1.0);
//=================================================================//
const int M=21;
const int dimB=20;
//=================================================================//
const double wbeg=-10.0;
const double wend=10.0;
const int wpoints=2000;
const double wpas=(wend-wbeg)/wpoints;
const double eta=0.1*w0;
//=================================================================//
int iz;
double Erreur;
double wtab[wpoints],As[wpoints],Ag[wpoints],Agk0[wpoints];
double Apol[wpoints];
double W[2][wpoints],S[2][wpoints],Sold[2][wpoints];
double A[M][wpoints];
//=================================================================//
int Inear(double a)
{
double a1=floor(a);
double a2=ceil(a);
if(fabs(a1-a)<fabs(a2-a)) return int(a1);
else return int(a2);
}
//=================================================================//
void initS()
{
for(int iw=0;iw<wpoints;iw++)
{
Sold[0][iw]=0.0;
Sold[1][iw]=0.0;
}
}
//=================================================================//
void mkW()
{
for(int iw=0;iw<wpoints;iw++)
{
double sumr=0.0;
double sumi=0.0;
for(int k=0;k<M;k++)
```

```cpp
{
double  ek=e0 −2.0∗t0∗cos (2.0∗ pi∗k/ double (M) ) ;
double  tampr=wtab [ iw]−ek−Sold [ 0 ] [ iw ] ;
double  tampi=eta −Sold [ 1 ] [ iw ] ;
double  denom=tampr∗tampr+tampi∗tampi ;
sumr+=tampr / denom ;
sumi −=tampi / denom ;
} // endor  k
sumr=sumr / double (M) ;
sumi=sumi / double (M) ;
double  denom=sumr∗sumr+sumi∗sumi ;
W[ 0 ] [ iw]=wtab [ iw]−e0−Sold [ 0 ] [ iw]−sumr / denom ;
W[ 1 ] [ iw]=eta −Sold [ 1 ] [ iw]+sumi / denom ;
} // endfor  iw
}
//===============================================================//
void  mkW2 ( )
{
for ( int  iw =0;iw<wpoints ; iw++)
{
double  sumr =0.0;
double  sumi =0.0;
for ( int  n=M−2;n >0;n−−)
{
double  tampr=wtab [ iw]−e0−Sold [ 0 ] [ iw]−sumr ;
double  tampi=eta −Sold [ 1 ] [ iw]−sumi ;
double  denom=tampr∗tampr+tampi∗tampi ;
sumr=t0 ∗t0 ∗tampr / denom ;
sumi=−t0 ∗t0 ∗tampi / denom ;
} // endor  n
double  tampr=wtab [ iw]−e0−Sold [ 0 ] [ iw]−sumr ;
double  tampi=eta −Sold [ 1 ] [ iw]−sumi ;
double  denom=tampr∗tampr+tampi∗tampi ;
W[ 0 ] [ iw]=2.0∗ t0 ∗t0 ∗tampr / denom ;
W[ 1 ] [ iw]=−2.0∗ t0 ∗t0 ∗tampi / denom ;
} // endfor  iw
}
//===============================================================//
void  saveAw ( )
{
FILE ∗fp ;
fp=fopen ( "Aw0. dat" ,"w" ) ;
for ( int  iw =0;iw<wpoints ; iw++)
{
double  tamp=−2.0∗W[ 1 ] [ iw ] ;
fprintf ( fp , "%18.18 f _%18.18 f _%18.18 f \n" , wtab [ iw ] ,W[ 0 ] [ iw ] , tamp ) ;
}
fclose ( fp ) ;
}
//===============================================================//
void  saveAs ( )
{
```

```cpp
FILE *fp;
fp=fopen("As.dat","w");
for(int iw=0;iw<wpoints;iw++)
{
fprintf(fp,"%18.18f_%18.18f\n",wtab[iw],As[iw]);
}
fclose(fp);
}
//==========================================================================//
void saveAg()
{
FILE *fp;
fp=fopen("Ag.dat","w");
for(int iw=0;iw<wpoints;iw++)
{
fprintf(fp,"%18.18f_%18.18f\n",wtab[iw],Ag[iw]);
}
fclose(fp);
}
//==========================================================================//
void saveAgk0()
{
FILE *fp;
fp=fopen("Agk0.dat","w");
for(int iw=0;iw<wpoints;iw++)
{
fprintf(fp,"%18.18f_%18.18f\n",wtab[iw],Agk0[iw]);
}
fclose(fp);
}
//==========================================================================//
void mkS()
{
for(int iw=0;iw<wpoints;iw++)
{
double sr=0.0;
double si=0.0;
for(int n=dimB;n>0;n--)
{
int i2=Inear(iw*1.0-n*w0/wpas);
double tampr=wtab[iw]-e0-n*w0-sr;
double tampi=eta-si;
if((i2>=0)&&(i2<wpoints)){ tampr-=W[0][i2]; tampi-=W[1][i2];}
double denom=tampr*tampr+tampi*tampi;
sr=n*g0*g0*w0*w0*tampr/denom;
si=-n*g0*g0*w0*w0*tampi/denom;
}//end n
S[0][iw]=sr;
S[1][iw]=si;
As[iw]=-2.0*si;
double tampr=wtab[iw]-e0-sr-W[0][iw];
double tampi=eta-si-W[1][iw];
```

```cpp
Ag[iw]=2.0*tampi/(tampr*tampr+tampi*tampi);
tampr=wtab[iw]-e0+2.0*t0-sr;
tampi=eta-si;
Agk0[iw]=2.0*tampi/(tampr*tampr+tampi*tampi);

}//end iw
}
//================================================================//
void mkError()
{
Erreur=0.0;
for(int iw=0;iw<wpoints;iw++)
{
Erreur+=fabs(S[0][iw]-Sold[0][iw])+fabs(S[1][iw]-Sold[1][iw]);
Sold[0][iw]=0.5*Sold[0][iw]+0.5*S[0][iw];
Sold[1][iw]=0.5*Sold[1][iw]+0.5*S[1][iw];
}
}
//================================================================//
void mkGK()
{
for(int k=0;k<M;k++)
{
double ek=e0-2.0*t0*cos(2.0*pi*k/(1.0*M));
printf("k=%d_ek=%lf\n",k,ek);
for(int iw=0;iw<wpoints;iw++)
{
double tampr=wtab[iw]-ek-S[0][iw];
double tampi=eta-S[1][iw];
double denom=tampr*tampr+tampi*tampi;
A[k][iw]=2.0*tampi/denom;
}//endfor iw
}//endfor k
}
//================================================================//
void saveGK()
{
double dec=2.0;
FILE *fp;
fp=fopen("A00.dat","w");
for(int iw=0;iw<wpoints;iw++)
fprintf(fp,"%18.18f_%18.18f\n",wtab[iw],A[0][iw]);
fclose(fp);
fp=fopen("A01.dat","w");
for(int iw=0;iw<wpoints;iw++)
fprintf(fp,"%18.18f_%18.18f\n",wtab[iw],1.0*dec+A[1][iw]);
fclose(fp);
fp=fopen("A02.dat","w");
for(int iw=0;iw<wpoints;iw++)
fprintf(fp,"%18.18f_%18.18f\n",wtab[iw],2.0*dec+A[2][iw]);
fclose(fp);
fp=fopen("A03.dat","w");
```

```cpp
for(int iw=0;iw<wpoints;iw++)
fprintf(fp,"%18.18f_%18.18f\n",wtab[iw],3.0*dec+A[3][iw]);
fclose(fp);
fp=fopen("A04.dat","w");
for(int iw=0;iw<wpoints;iw++)
fprintf(fp,"%18.18f_%18.18f\n",wtab[iw],4.0*dec+A[4][iw]);
fclose(fp);
fp=fopen("A05.dat","w");
for(int iw=0;iw<wpoints;iw++)
fprintf(fp,"%18.18f_%18.18f\n",wtab[iw],5.0*dec+A[5][iw]);
fclose(fp);
fp=fopen("A06.dat","w");
for(int iw=0;iw<wpoints;iw++)
fprintf(fp,"%18.18f_%18.18f\n",wtab[iw],6.0*dec+A[6][iw]);
fclose(fp);
fp=fopen("A07.dat","w");
for(int iw=0;iw<wpoints;iw++)
fprintf(fp,"%18.18f_%18.18f\n",wtab[iw],7.0*dec+A[7][iw]);
fclose(fp);
fp=fopen("A08.dat","w");
for(int iw=0;iw<wpoints;iw++)
fprintf(fp,"%18.18f_%18.18f\n",wtab[iw],8.0*dec+A[8][iw]);
fclose(fp);
fp=fopen("A09.dat","w");
for(int iw=0;iw<wpoints;iw++)
fprintf(fp,"%18.18f_%18.18f\n",wtab[iw],9.0*dec+A[9][iw]);
fclose(fp);
fp=fopen("A10.dat","w");
for(int iw=0;iw<wpoints;iw++)
fprintf(fp,"%18.18f_%18.18f\n",wtab[iw],10.0*dec+A[10][iw]);
fclose(fp);

}
//==================================================================//
main()
{
for(int iw=0;iw<wpoints;iw++)  wtab[iw]=wbeg+iw*wpas;
initS();
do
{
mkW2();
mkS();
mkError();
printf("Erreur=%18.18f\n",Erreur);
}while(Erreur>10e-6);
saveAw();
saveAs();
saveAg();
saveAgk0();
mkGK();
saveGK();
}
```

//===//

disp11e.cc

```cpp
//=================================================================//
// disp11e . cc
//=================================================================//
#include <cmath>
#include <cstdio>
#include <cstdlib>
#include "ludcmp.h"
#include "lubksb.h"
#include "mprove.h"
//=================================================================//
const double e0 =0.0;
const double w0 =1.0;
const double t0 =1.0;
const double g0 =1.0;
const double pi =acos ( -1.0);
//=================================================================//
const int M=21;
const int dimB =20;
//=================================================================//
const double wbeg = -10.0;
const double wend =10.0;
const int wpoints =2000;
const double wpas =(wend-wbeg )/ wpoints ;
const double eta =0.1*w0;
//=================================================================//
double Erreur ;
double wtab[ wpoints ] , As[ wpoints ] , Ag[ wpoints ] , Agk0[ wpoints ] ;
double Apol[ wpoints ] ;
double W[2][ wpoints ] , S[2][ wpoints ] , Sold [2][ wpoints ] ;
double WNr[M][ wpoints ] , WNi[M][ wpoints ] ;
double GNr[M][ wpoints ] , GNi[M][ wpoints ] ;
double GNNr[M][ wpoints ] , GNNi[M][ wpoints ] ;
double GKr[M][ wpoints ] , GKi[M][ wpoints ] ;
double AK[M][ wpoints ] ;
//=================================================================//
int Inear (double a)
{
double a1=floor (a );
double a2=ceil (a );
if ( fabs (a1-a)< fabs (a2-a)) return int (a1 );
else return int (a2 );
}
//=================================================================//
void initS ()
{
for ( int iw =0; iw<wpoints ; iw++)
{
Sold [0][ iw ]=0.0;
```

```cpp
Sold[1][iw]=0.0;
}
}
//===============================================================//
void mkW()
{
for(int iw=0;iw<wpoints;iw++)
{
double sumr=0.0;
double sumi=0.0;
for(int k=0;k<M;k++)
{
double ek=e0-2.0*t0*cos(2.0*pi*k/double(M));
double tampr=wtab[iw]-ek-Sold[0][iw];
double tampi=eta-Sold[1][iw];
double denom=tampr*tampr+tampi*tampi;
sumr+=tampr/denom;
sumi-=tampi/denom;
}//endor k
sumr=sumr/double(M);
sumi=sumi/double(M);
double denom=sumr*sumr+sumi*sumi;
W[0][iw]=wtab[iw]-e0-Sold[0][iw]-sumr/denom;
W[1][iw]=eta-Sold[1][iw]+sumi/denom;
}//endfor iw
}
//===============================================================//
void mkW2()
{
for(int iw=0;iw<wpoints;iw++)
{
double sumr=0.0;
double sumi=0.0;
for(int n=M-2;n>0;n--)
{
double tampr=wtab[iw]-e0-Sold[0][iw]-sumr;
double tampi=eta-Sold[1][iw]-sumi;
double denom=tampr*tampr+tampi*tampi;
sumr=t0*t0*tampr/denom;
sumi=-t0*t0*tampi/denom;
}//endor n
double tampr=wtab[iw]-e0-Sold[0][iw]-sumr;
double tampi=eta-Sold[1][iw]-sumi;
double denom=tampr*tampr+tampi*tampi;
W[0][iw]=2.0*t0*t0*tampr/denom;
W[1][iw]=-2.0*t0*t0*tampi/denom;
}//endfor iw
}
//===============================================================//
void saveAw()
{
FILE *fp;
```

```cpp
fp=fopen("Aw.dat","w");
for(int iw=0;iw<wpoints;iw++)
{
double tamp=-2.0*W[1][iw];
fprintf(fp,"%18.18f_%18.18f\n",wtab[iw],tamp);
}
fclose(fp);
}
//================================================================//
void saveAw0()
{
FILE *fp;
fp=fopen("Aw0.dat","w");
for(int iw=0;iw<wpoints;iw++)
{
double tamp=-2.0*WNi[0][iw];
fprintf(fp,"%18.18f_%18.18f\n",wtab[iw],tamp);
}
fclose(fp);
}
//================================================================//
void saveAg0()
{
FILE *fp;
fp=fopen("Ag0.dat","w");
for(int iw=0;iw<wpoints;iw++)
{
double tamp=-2.0*GNi[0][iw];
fprintf(fp,"%18.18f_%18.18f\n",wtab[iw],tamp);
}
fclose(fp);
}
//================================================================//

void saveAs()
{
FILE *fp;
fp=fopen("As.dat","w");
for(int iw=0;iw<wpoints;iw++)
{
fprintf(fp,"%18.18f_%18.18f\n",wtab[iw],As[iw]);
}
fclose(fp);
}
//================================================================//
void saveAg()
{
FILE *fp;
fp=fopen("Ag.dat","w");
for(int iw=0;iw<wpoints;iw++)
{
fprintf(fp,"%18.18f_%18.18f\n",wtab[iw],Ag[iw]);
```

```cpp
}
fclose(fp);
}
//=================================================================//
void saveAgk0()
{
FILE *fp;
fp=fopen("Agk0.dat","w");
for(int iw=0;iw<wpoints;iw++)
{
fprintf(fp,"%18.18f_%18.18f\n",wtab[iw],Agk0[iw]);
}
fclose(fp);
}
//=================================================================//
void mkS()
{
for(int iw=0;iw<wpoints;iw++)
{
double sr=0.0;
double si=0.0;
for(int n=dimB;n>0;n--)
{
int i2=Inear(iw*1.0-n*w0/wpas);
double tampr=wtab[iw]-e0-n*w0-sr;
double tampi=eta-si;
if((i2>=0)&&(i2<wpoints)){ tampr-=W[0][i2]; tampi-=W[1][i2];}
double denom=tampr*tampr+tampi*tampi;
sr=n*g0*g0*w0*w0*tampr/denom;
si=-n*g0*g0*w0*w0*tampi/denom;
}//end n
S[0][iw]=sr;
S[1][iw]=si;
As[iw]=-2.0*si;
double tampr=wtab[iw]-e0-sr-W[0][iw];
double tampi=eta-si-W[1][iw];
Ag[iw]=2.0*tampi/(tampr*tampr+tampi*tampi);
tampr=wtab[iw]-e0+2.0*t0-sr;
tampi=eta-si;
Agk0[iw]=2.0*tampi/(tampr*tampr+tampi*tampi);

}//end iw
}
//=================================================================//
void mkError()
{
Erreur=0.0;
for(int iw=0;iw<wpoints;iw++)
{
Erreur+=fabs(S[0][iw]-Sold[0][iw])+fabs(S[1][iw]-Sold[1][iw]);
Sold[0][iw]=0.5*Sold[0][iw]+0.5*S[0][iw];
Sold[1][iw]=0.5*Sold[1][iw]+0.5*S[1][iw];
```

```cpp
}
}
//=================================================================//
void mkWN()
{
for(int iw=0;iw<wpoints;iw++)
{
WNr[M-1][iw]=0.0;
WNi[M-1][iw]=0.0;
}
for(int n=M-2;n>0;n--)
{
for(int iw=0;iw<wpoints;iw++)
{
double tampr=wtab[iw]-e0-S[0][iw]-WNr[n+1][iw];
double tampi=eta-S[1][iw]-WNi[n+1][iw];
double denom=tampr*tampr+tampi*tampi;
WNr[n][iw]=t0*t0*tampr/denom;
WNi[n][iw]=-t0*t0*tampi/denom;
}//enfor iw
}//endfor n
for(int iw=0;iw<wpoints;iw++)
{
double tampr=wtab[iw]-e0-S[0][iw]-WNr[1][iw];
double tampi=eta-S[1][iw]-WNi[1][iw];
double denom=tampr*tampr+tampi*tampi;
WNr[0][iw]=2.0*t0*t0*tampr/denom;
WNi[0][iw]=-2.0*t0*t0*tampi/denom;
}//enfor iw

}
//=================================================================//
void mkGN()
{
for(int n=0;n<M;n++)
{
for(int iw=0;iw<wpoints;iw++)
{
double tampr=wtab[iw]-e0-S[0][iw]-WNr[n][iw];
double tampi=eta-S[1][iw]-WNi[n][iw];
double denom=tampr*tampr+tampi*tampi;
GNr[n][iw]=tampr/denom;
GNi[n][iw]=-tampi/denom;
}
}
}
//=================================================================//
void mkGNN()
{
for(int iw=0;iw<wpoints;iw++)
{
GNNr[0][iw]=GNr[0][iw];
```

```cpp
GNNi[0][iw]=GNi[0][iw];
}
for(int  n=1;n<M;n++)
{
for(int  iw=0;iw<wpoints;iw++)
{
double  tampr=GNNr[n-1][iw]*GNr[n][iw]-GNNi[n-1][iw]*GNi[n][iw];
double  tampi=GNNr[n-1][iw]*GNi[n][iw]+GNNi[n-1][iw]*GNr[n][iw];
GNNr[n][iw]=-t0*tampr;
GNNi[n][iw]=-t0*tampi;
}//endfor  iw
}//endfor  n
}
//==================================================================//
void  mkGK()
{
for(int  k=0;k<M;k++)
{
for(int  iw=0;iw<wpoints;iw++)
{
GKr[k][iw]=GNNr[0][iw];
GKi[k][iw]=GNNi[0][iw];
for(int  n=1;n<(M+1)/2;n++)
{
GKr[k][iw]+=2.0*cos(2.0*pi*k*1.0*n*1.0/double(M))*GNNr[n][iw];
GKi[k][iw]+=2.0*cos(2.0*pi*k*1.0*n*1.0/double(M))*GNNi[n][iw];
}//endfor  n
AK[k][iw]=-2.0*GKi[k][iw];
}//endor  iw
}//enfor  k
}
//==================================================================//
void  saveGK()
{
double  dec=2.0;
FILE *fp;
fp=fopen("A00.dat","w");
for(int  iw=0;iw<wpoints;iw++)
fprintf(fp,"%18.18f_%18.18f\n",wtab[iw],AK[0][iw]);
fclose(fp);
fp=fopen("A01.dat","w");
for(int  iw=0;iw<wpoints;iw++)
fprintf(fp,"%18.18f_%18.18f\n",wtab[iw],1.0*dec+AK[1][iw]);
fclose(fp);
fp=fopen("A02.dat","w");
for(int  iw=0;iw<wpoints;iw++)
fprintf(fp,"%18.18f_%18.18f\n",wtab[iw],2.0*dec+AK[2][iw]);
fclose(fp);
fp=fopen("A03.dat","w");
for(int  iw=0;iw<wpoints;iw++)
fprintf(fp,"%18.18f_%18.18f\n",wtab[iw],3.0*dec+AK[3][iw]);
fclose(fp);
```

```cpp
fp=fopen("A04.dat","w");
for(int iw=0;iw<wpoints;iw++)
fprintf(fp,"%18.18f_%18.18f\n",wtab[iw],4.0*dec+AK[4][iw]);
fclose(fp);
fp=fopen("A05.dat","w");
for(int iw=0;iw<wpoints;iw++)
fprintf(fp,"%18.18f_%18.18f\n",wtab[iw],5.0*dec+AK[5][iw]);
fclose(fp);
fp=fopen("A06.dat","w");
for(int iw=0;iw<wpoints;iw++)
fprintf(fp,"%18.18f_%18.18f\n",wtab[iw],6.0*dec+AK[6][iw]);
fclose(fp);
fp=fopen("A07.dat","w");
for(int iw=0;iw<wpoints;iw++)
fprintf(fp,"%18.18f_%18.18f\n",wtab[iw],7.0*dec+AK[7][iw]);
fclose(fp);
fp=fopen("A08.dat","w");
for(int iw=0;iw<wpoints;iw++)
fprintf(fp,"%18.18f_%18.18f\n",wtab[iw],8.0*dec+AK[8][iw]);
fclose(fp);
fp=fopen("A09.dat","w");
for(int iw=0;iw<wpoints;iw++)
fprintf(fp,"%18.18f_%18.18f\n",wtab[iw],9.0*dec+AK[9][iw]);
fclose(fp);
fp=fopen("A10.dat","w");
for(int iw=0;iw<wpoints;iw++)
fprintf(fp,"%18.18f_%18.18f\n",wtab[iw],10.0*dec+AK[10][iw]);
fclose(fp);
}
//================================================================//
main()
{
for(int iw=0;iw<wpoints;iw++)  wtab[iw]=wbeg+iw*wpas;
initS();
do
{
mkW2();
mkS();
mkError();
printf("Erreur=%18.18f\n",Erreur);
} while(Erreur>10e-6);
saveAw();
saveAs();
saveAg();
saveAgk0();

mkWN();
saveAw0();
mkGN();
saveAg0();
mkGNN();
mkGK();
```

```cpp
saveGK ();
}
//===========================================================================//
```

disp11p.cc

```cpp
//===========================================================================//
// disp11p.cc
//===========================================================================//
#include  <cmath>
#include  <cstdio>
#include  <cstdlib>
#include  "ludcmp.h"
#include  "lubksb.h"
#include  "mprove.h"
//===========================================================================//
const  double  e0=0.0;
const  double  w0=1.0;
const  double  t0=1.0;
const  double  g0=1.0;
const  double  pi=acos(-1.0);
//===========================================================================//
const  int  M=41;
const  int  dimB=20;
//===========================================================================//
const  double  wbeg=-10.0;
const  double  wend=10.0;
const  int  wpoints=2000;
const  double  wpas=(wend-wbeg)/wpoints;
const  double  eta=0.1*w0;
//===========================================================================//
int  iz,im;
double  Erreur;
double  wtab[wpoints],As[wpoints],Ag[wpoints],Agk0[wpoints];
double  Apol[wpoints];
double  W[2][wpoints],S[2][wpoints],Sold[2][wpoints];
double  WNr[M][wpoints],WNi[M][wpoints];
double  U[dimB];
double  Kr[dimB][dimB],Ki[dimB][dimB];
double  GNr[M][dimB][dimB],GNi[M][dimB][dimB];
double  GNNr[M][wpoints],GNNi[M][wpoints];
double  GKr[M][wpoints],GKi[M][wpoints];
double  AK[M][wpoints];
//===========================================================================//
int  Inear(double  a)
{
double  a1=floor(a);
double  a2=ceil(a);
if(fabs(a1-a)<fabs(a2-a))  return  int(a1);
else  return  int(a2);
}
//===========================================================================//
```

```cpp
void  initS ()
{
for ( int  iw =0;iw<wpoints ; iw ++)
{
Sold [ 0 ] [ iw ]=0.0;
Sold [ 1 ] [ iw ]=0.0;
}
}
//=================================================================//
void  mkW ()
{
for ( int  iw =0;iw<wpoints ; iw ++)
{
double  sumr =0.0;
double  sumi =0.0;
for ( int  k =0;k<M; k ++)
{
double  ek=e0 −2.0* t0 *cos ( 2.0 * pi *k / double (M));
double  tampr=wtab [ iw ]−ek−Sold [ 0 ] [ iw ];
double  tampi=eta −Sold [ 1 ] [ iw ];
double  denom=tampr *tampr+tampi *tampi ;
sumr+=tampr / denom ;
sumi −=tampi / denom ;
} // endor  k
sumr=sumr / double (M);
sumi=sumi / double (M);
double  denom=sumr *sumr+sumi *sumi ;
W[ 0 ] [ iw ]= wtab [ iw ]−e0−Sold [ 0 ] [ iw ]−sumr / denom ;
W[ 1 ] [ iw ]= eta −Sold [ 1 ] [ iw ]+sumi / denom ;
} // endfor  iw
}
//=================================================================//
void  mkW2 ()
{
for ( int  iw =0;iw<wpoints ; iw ++)
{
double  sumr =0.0;
double  sumi =0.0;
for ( int  n=M−2;n >0;n −−)
{
double  tampr=wtab [ iw ]−e0−Sold [ 0 ] [ iw ]−sumr ;
double  tampi=eta −Sold [ 1 ] [ iw ]−sumi ;
double  denom=tampr *tampr+tampi *tampi ;
sumr=t0 * t0 *tampr / denom ;
sumi=−t0 * t0 *tampi / denom ;
} // endor  n
double  tampr=wtab [ iw ]−e0−Sold [ 0 ] [ iw ]−sumr ;
double  tampi=eta −Sold [ 1 ] [ iw ]−sumi ;
double  denom=tampr *tampr+tampi *tampi ;
W[ 0 ] [ iw ]=2.0 * t0 * t0 *tampr / denom ;
W[ 1 ] [ iw ]=−2.0* t0 * t0 *tampi / denom ;
} // endfor  iw
```

```cpp
}
//================================================================//
void saveAw()
{
FILE *fp;
fp=fopen("Aw.dat","w");
for(int iw=0;iw<wpoints;iw++)
{
double tamp=-2.0*W[1][iw];
fprintf(fp,"%18.18f %18.18f\n",wtab[iw],tamp);
}
fclose(fp);
}
//================================================================//
void saveAw0()
{
FILE *fp;
fp=fopen("Aw0.dat","w");
for(int iw=0;iw<wpoints;iw++)
{
double tamp=-2.0*WNi[0][iw];
fprintf(fp,"%18.18f %18.18f\n",wtab[iw],tamp);
}
fclose(fp);
}
//================================================================//
void saveAs()
{
FILE *fp;
fp=fopen("As.dat","w");
for(int iw=0;iw<wpoints;iw++)
{
fprintf(fp,"%18.18f %18.18f\n",wtab[iw],As[iw]);
}
fclose(fp);
}
//================================================================//
void saveAg()
{
FILE *fp;
fp=fopen("Ag.dat","w");
for(int iw=0;iw<wpoints;iw++)
{
fprintf(fp,"%18.18f %18.18f\n",wtab[iw],Ag[iw]);
}
fclose(fp);
}
//================================================================//
void saveAgk0()
{
FILE *fp;
fp=fopen("Agk0.dat","w");
```

```cpp
for(int iw=0;iw<wpoints;iw++)
{
fprintf(fp,"%18.18f_%18.18f\n",wtab[iw],Agk0[iw]);
}
fclose(fp);
}
//================================================================//
void mkS()
{
for(int iw=0;iw<wpoints;iw++)
{
double sr=0.0;
double si=0.0;
for(int n=dimB;n>0;n--)
{
int i2=Inear(iw*1.0-n*w0/wpas);
double tampr=wtab[iw]-e0-n*w0-sr;
double tampi=eta-si;
if((i2>=0)&&(i2<wpoints)){tampr-=W[0][i2];tampi-=W[1][i2];}
double denom=tampr*tampr+tampi*tampi;
sr=n*g0*g0*w0*w0*tampr/denom;
si=-n*g0*g0*w0*w0*tampi/denom;
}//end n
S[0][iw]=sr;
S[1][iw]=si;
As[iw]=-2.0*si;
double tampr=wtab[iw]-e0-sr-W[0][iw];
double tampi=eta-si-W[1][iw];
Ag[iw]=2.0*tampi/(tampr*tampr+tampi*tampi);
tampr=wtab[iw]-e0+2.0*t0-sr;
tampi=eta-si;
Agk0[iw]=2.0*tampi/(tampr*tampr+tampi*tampi);

}//end iw
}
//================================================================//
void mkError()
{
Erreur=0.0;
for(int iw=0;iw<wpoints;iw++)
{
Erreur+=fabs(S[0][iw]-Sold[0][iw])+fabs(S[1][iw]-Sold[1][iw]);
Sold[0][iw]=0.5*Sold[0][iw]+0.5*S[0][iw];
Sold[1][iw]=0.5*Sold[1][iw]+0.5*S[1][iw];
}
}
//================================================================//
void mkWN()
{
for(int iw=0;iw<wpoints;iw++)
{
WNr[M-1][iw]=0.0;
```

```cpp
WNi[M-1][iw]=0.0;
}
for(int  n=M-2;n>0;n--)
{
for(int  iw=0;iw<wpoints;iw++)
{
double  tampr=wtab[iw]-e0-S[0][iw]-WNr[n+1][iw];
double  tampi=eta-S[1][iw]-WNi[n+1][iw];
double  denom=tampr*tampr+tampi*tampi;
WNr[n][iw]=t0*t0*tampr/denom;
WNi[n][iw]=-t0*t0*tampi/denom;
}//enfor  iw
}//endfor  n
for(int  iw=0;iw<wpoints;iw++)
{
double  tampr=wtab[iw]-e0-S[0][iw]-WNr[1][iw];
double  tampi=eta-S[1][iw]-WNi[1][iw];
double  denom=tampr*tampr+tampi*tampi;
WNr[0][iw]=2.0*t0*t0*tampr/denom;
WNi[0][iw]=-2.0*t0*t0*tampi/denom;
}//enfor  iw
}
//=================================================================//
void  mkKloc()
{
for(int  i=0;i<dimB;i++)  for(int  j=0;j<dimB;j++)  Kr[i][j]=0.0;
for(int  i=0;i<dimB;i++)  for(int  j=0;j<dimB;j++)  Ki[i][j]=0.0;
for(int  n=0;n<dimB;n++)  Kr[n][n]=wtab[iz]-e0-n*w0;
for(int  n=0;n<dimB;n++)  Ki[n][n]=eta;
for(int  i=1;i<dimB;i++)  Kr[i-1][i]=Kr[i][i-1]=g0*w0*sqrt(1.0*i);
Kr[0][0]  -=WNr[im][iz];
Ki[0][0]  -=WNi[im][iz];
for(int  n=1;n<dimB;n++)
{
int  i2=Inear(iz*1.0-n*w0/wpas);
if(i2>=0){Kr[n][n]-=W[0][i2];Ki[n][n]-=W[1][i2];}
}
}
//=================================================================//
void  mkGloc()
{
double  **g_mat;g_mat=new  double*[2*dimB];
for(int  i=0;i<2*dimB;i++)  g_mat[i]=new  double[2*dimB];
double  **glu_mat;glu_mat=new  double*[2*dimB];
for(int  i=0;i<2*dimB;i++)  glu_mat[i]=new  double[2*dimB];
double  *x_vec=new  double[2*dimB];
double  *b_vec=new  double[2*dimB];
int  *perm_vec=new  int[2*dimB];
double  dperm;
double  err;

for(int  i=0;i<dimB;i++)
```

```cpp
{
for(int   j=0;j<dimB;j++)
{
g_mat[i][j]=Kr[i][j];
g_mat[i][j+dimB]=-Ki[i][j];
g_mat[i+dimB][j]=Ki[i][j];
g_mat[i+dimB][j+dimB]=Kr[i][j];
}
}

for(int  i=0;i<2*dimB;i++)  for(int   j=0;j<2*dimB;j++)
glu_mat[i][j]=g_mat[i][j];

ludcmp(glu_mat,2*dimB,perm_vec,dperm);

for(int  n=0;n<dimB;n++)
{
for(int  i=0;i<2*dimB;i++)  x_vec[i]=0.0;x_vec[n]=1.0;
lubksb(glu_mat,2*dimB,perm_vec,x_vec);
for(int  i=0;i<2*dimB;i++)  b_vec[i]=0.0;b_vec[n]=1.0;
do
{
mprove(g_mat,glu_mat,2*dimB,perm_vec,b_vec,x_vec,err);
}while(err>10e-12);
for(int  i=0;i<dimB;i++) GNr[im][i][n]=x_vec[i];
for(int  i=0;i<dimB;i++) GNi[im][i][n]=x_vec[i+dimB];
}

for(int  i=0;i<2*dimB;i++)  delete [] g_mat[i];delete [] g_mat;
for(int  i=0;i<2*dimB;i++)  delete [] glu_mat[i];delete [] glu_mat;
delete [] x_vec;
delete [] b_vec;
delete [] perm_vec;
}
//===============================================================//
void mkU()
{
double tamp=exp(-0.5*g0*g0);
U[0]=tamp;
for(int  n=1;n<dimB;n++)
{
tamp=tamp*g0/sqrt(1.0*n);
U[n]=tamp;
}
}
//===============================================================//
void mkG00()
{
GNNr[0][iz]=0.0;
GNNi[0][iz]=0.0;
for(int  i=0;i<dimB;i++)
{
```

```cpp
for(int  j =0; j<dimB ; j ++)
{
GNNr[0][ iz ]+=U[ i ]*U[ j ]*GNr[0][ i ][ j ];
GNNi[0][ iz ]+=U[ i ]*U[ j ]*GNi[0][ i ][ j ];
}
}
}
//================================================================//
void  mkG01 ()
{
double  glr =0.0;
double  gli =0.0;
for(int  i =0; i<dimB ; i ++)
{
glr+=U[ i ]*GNr[0][ i ][0];
gli+=U[ i ]*GNi[0][ i ][0];
}
double  grr =0.0;
double  gri =0.0;
for(int  i =0; i<dimB ; i ++)
{
grr+=GNr[ 1 ][0][ i ]*U[ i ];
gri+=GNi[ 1 ][0][ i ]*U[ i ];
}
double  tampr=glr*grr−gli*gri ;
double  tampi=glr*gri+gli*grr ;
GNNr[ 1 ][ iz ]=− t0 *tampr ;
GNNi[ 1 ][ iz ]=− t0 *tampi ;
}
//================================================================//
void  mkG0m ()
{
double  glr =0.0;
double  gli =0.0;
for(int  i =0; i<dimB ; i ++)
{
glr+=U[ i ]*GNr[0][ i ][0];
gli+=U[ i ]*GNi[0][ i ][0];
}

double  grr =0.0;
double  gri =0.0;
for(int  i =0; i<dimB ; i ++)
{
grr+=GNr[ im ][0][ i ]*U[ i ];
gri+=GNi[ im ][0][ i ]*U[ i ];
}
grr=−t0 *grr ;
gri=−t0 *gri ;

double  gmr=−t0 *GNr[ 1 ][0][0];
double  gmi=−t0 *GNi[ 1 ][0][0];
```

```cpp
for(int n=2;n<im;n++)
{
double tamp1=GNr[n][0][0];
double tamp2=GNi[n][0][0];
double tampr=gmr*tamp1-gmi*tamp2;
double tampi=gmr*tamp2+gmi*tamp1;
gmr=-t0*tampr;
gmi=-t0*tampi;
}

double tampr=glr*gmr-gli*gmi;
double tampi=glr*gmi+gli*gmr;
GNNr[im][iz]=tampr*grr-tampi*gri;
GNNi[im][iz]=tampr*gri+tampi*grr;
}
//================================================================//
void mkGK()
{
for(int k=0;k<M;k++)
{
for(int iw=0;iw<wpoints;iw++)
{
GKr[k][iw]=GNNr[0][iw];
GKi[k][iw]=GNNi[0][iw];
for(int n=1;n<(M+1)/2;n++)
{
GKr[k][iw]+=2.0*cos(2.0*pi*k*1.0*n*1.0/double(M))*GNNr[n][iw];
GKi[k][iw]+=2.0*cos(2.0*pi*k*1.0*n*1.0/double(M))*GNNi[n][iw];
}//endfor n
AK[k][iw]=-2.0*GKi[k][iw];
}//endor iw
}//enfor k
}
//================================================================//
void saveGK()
{
double dec=2.0;
FILE *fp;
fp=fopen("A00.dat","w");
for(int iw=0;iw<wpoints;iw++)
fprintf(fp,"%18.18f_%18.18f\n",wtab[iw],AK[0][iw]);
fclose(fp);
fp=fopen("A01.dat","w");
for(int iw=0;iw<wpoints;iw++)
fprintf(fp,"%18.18f_%18.18f\n",wtab[iw],1.0*dec+AK[1][iw]);
fclose(fp);
fp=fopen("A02.dat","w");
for(int iw=0;iw<wpoints;iw++)
fprintf(fp,"%18.18f_%18.18f\n",wtab[iw],2.0*dec+AK[2][iw]);
fclose(fp);
fp=fopen("A03.dat","w");
for(int iw=0;iw<wpoints;iw++)
```

```cpp
fprintf(fp,"%18.18f_%18.18f\n",wtab[iw],3.0*dec+AK[3][iw]);
fclose(fp);
fp=fopen("A04.dat","w");
for(int iw=0;iw<wpoints;iw++)
fprintf(fp,"%18.18f_%18.18f\n",wtab[iw],4.0*dec+AK[4][iw]);
fclose(fp);
fp=fopen("A05.dat","w");
for(int iw=0;iw<wpoints;iw++)
fprintf(fp,"%18.18f_%18.18f\n",wtab[iw],5.0*dec+AK[5][iw]);
fclose(fp);
fp=fopen("A06.dat","w");
for(int iw=0;iw<wpoints;iw++)
fprintf(fp,"%18.18f_%18.18f\n",wtab[iw],6.0*dec+AK[6][iw]);
fclose(fp);
fp=fopen("A07.dat","w");
for(int iw=0;iw<wpoints;iw++)
fprintf(fp,"%18.18f_%18.18f\n",wtab[iw],7.0*dec+AK[7][iw]);
fclose(fp);
fp=fopen("A08.dat","w");
for(int iw=0;iw<wpoints;iw++)
fprintf(fp,"%18.18f_%18.18f\n",wtab[iw],8.0*dec+AK[8][iw]);
fclose(fp);
fp=fopen("A09.dat","w");
for(int iw=0;iw<wpoints;iw++)
fprintf(fp,"%18.18f_%18.18f\n",wtab[iw],9.0*dec+AK[9][iw]);
fclose(fp);
fp=fopen("A10.dat","w");
for(int iw=0;iw<wpoints;iw++)
fprintf(fp,"%18.18f_%18.18f\n",wtab[iw],10.0*dec+AK[10][iw]);
fclose(fp);
}
//==============================================================//
main()
{
for(int iw=0;iw<wpoints;iw++)  wtab[iw]=wbeg+iw*wpas;
initS();
do
{
mkW2();
mkS();
mkError();
printf("Erreur=%18.18f\n",Erreur);
}while(Erreur>10e-12);
saveAw();
saveAs();
saveAg();
saveAgk0();

mkWN();
saveAw0();
mkU();
for(iz=0;iz<wpoints;iz++)
```

```cpp
{
for ( im = 0 ; im<M ; im++){ mkKloc ( ) ; mkGloc ( ) ; }
mkG00 ( ) ;
mkG01 ( ) ;
for ( im = 2 ; im<M ; im++)  mkG0m ( ) ;
}
mkGK ( ) ;
saveGK ( ) ;
}
//================================================================//
```

exact2p.cc

```cpp
//================================================================//
// exact2p . cc
//================================================================//
// Holstein  2  sites
// Local  Green's  function  P11(z)
//================================================================//
#include  <cmath>
#include  <cstdio>
#include  <cstdlib>
#include  "jacobi .h"
#include  "eigsrt .h"
//================================================================//
const  double  e0 =0.0;
const  double  w0=1.0;
const  double  t0 =1.0;
const  double  g0=1.0;
const  int  dimB =25;
const  int  dimBB=dimB*dimB ;
const  int  dimH=2*dimBB ;
//================================================================//
int  NB1[dimBB] ,NB2[dimBB] ,NB[dimBB] ;
double  H11[dimBB][dimBB] ,H12[dimBB][dimBB] ;
double  H21[dimBB][dimBB] ,H22[dimBB][dimBB] ;
double  Ep11[dimBB] ,Vp11[dimBB][dimBB] ;
double  Ep22[dimBB] ,Vp22[dimBB][dimBB] ;
double  Hp11[dimBB][dimBB] ,Hp12[dimBB][dimBB] ;
double  Hp21[dimBB][dimBB] ,Hp22[dimBB][dimBB] ;
double  H[dimH][dimH] ;
double  Ep[dimH] ,V11[dimH] ,V12[dimH] ,Vss[dimH] ,Vaa[dimH] ;
//================================================================//
void  mkNB ()
{
for ( int  n1 =0;n1<dimB ; n1++)
{
for ( int  n2 =0;n2<dimB ; n2++)
{
NB1[ n1+dimB*n2]=n1 ;
NB2[ n1+dimB*n2 ]=n2 ;
NB[ n1+dimB*n2 ]=n1+n2 ;
```

```cpp
}
}
}
//=====================================================================//
void mkH11()
{
for(int m=0;m<dimBB;m++) for(int n=0;n<dimBB;n++) H11[m][n]=0.0;
for(int n=0;n<dimBB;n++) H11[n][n]=e0+w0*NB[n];
for(int m=0;m<dimBB;m++)
{
for(int n=0;n<dimBB;n++)
{
int tutu=0;
if(NB2[m]==NB2[n]) tutu+=1;
if(NB1[m]==(NB1[n]-1)) tutu+=1;
if(tutu==2)
{
H11[m][n]=-g0*w0*sqrt(1.0*NB1[n]);
H11[n][m]=-g0*w0*sqrt(1.0*NB1[n]);
}
}
}
}
//=====================================================================//
void mkH22()
{
for(int m=0;m<dimBB;m++) for(int n=0;n<dimBB;n++) H22[m][n]=0.0;
for(int n=0;n<dimBB;n++) H22[n][n]=e0+w0*NB[n];
for(int m=0;m<dimBB;m++)
{
for(int n=0;n<dimBB;n++)
{
int tutu=0;
if(NB1[m]==NB1[n]) tutu+=1;
if(NB2[m]==(NB2[n]-1)) tutu+=1;
if(tutu==2)
{
H22[m][n]=-g0*w0*sqrt(1.0*NB2[n]);
H22[n][m]=-g0*w0*sqrt(1.0*NB2[n]);
}
}
}
}
//=====================================================================//
void mkH12()
{
for(int m=0;m<dimBB;m++) for(int n=0;n<dimBB;n++) H12[m][n]=0.0;
for(int m=0;m<dimBB;m++) H12[m][m]=-t0;
}
//=====================================================================//
void mkH21()
{
```

```cpp
for(int m=0;m<dimBB;m++) for(int n=0;n<dimBB;n++) H21[m][n]=0.0;
for(int m=0;m<dimBB;m++) H21[m][m]=-t0;
}
//================================================================//
void mkH()
{
for(int m=0;m<dimBB;m++)
{
for(int n=0;n<dimBB;n++)
{
H[m][n]=Hp11[m][n];
H[m][n+dimBB]=Hp12[m][n];
H[m+dimBB][n]=Hp21[m][n];
H[m+dimBB][n+dimBB]=Hp22[m][n];
}
}
}
//================================================================//
void seeH11()
{
for(int i=0;i<dimBB;i++)
{
for(int j=0;j<dimBB;j++) printf("%2.2f_",H11[i][j]);
printf("\n");
}
}
//================================================================//
void diagH11()
{
double **h_mat;h_mat=new double *[dimBB];
for(int i=0;i<dimBB;i++) h_mat[i]=new double[dimBB];
double **v_mat;v_mat=new double *[dimBB];
for(int i=0;i<dimBB;i++) v_mat[i]=new double[dimBB];
double *d_vec=new double[dimBB];

for(int i=0;i<dimBB;i++) for(int j=0;j<dimBB;j++)
                              h_mat[i][j]=H11[i][j];
jacobi(h_mat,dimBB,d_vec,v_mat,10e-24);
eigsrt(d_vec,v_mat,dimBB);
for(int i=0;i<dimBB;i++) Ep11[i]=d_vec[i];
for(int i=0;i<dimBB;i++) for(int j=0;j<dimBB;j++)
                              Vp11[i][j]=v_mat[i][j];

for(int i=0;i<dimBB;i++) delete [] h_mat[i];delete [] h_mat;
for(int i=0;i<dimBB;i++) delete [] v_mat[i];delete [] v_mat;
delete [] d_vec;
}
//================================================================//
void diagH22()
{
double **h_mat;h_mat=new double *[dimBB];
for(int i=0;i<dimBB;i++) h_mat[i]=new double[dimBB];
```

```cpp
double **v_mat;v_mat=new double *[dimBB];
for(int i=0;i<dimBB;i++) v_mat[i]=new double[dimBB];
double *d_vec=new double[dimBB];

for(int i=0;i<dimBB;i++) for(int j=0;j<dimBB;j++)
                                h_mat[i][j]=H22[i][j];
jacobi(h_mat,dimBB,d_vec,v_mat,10e-24);
eigsrt(d_vec,v_mat,dimBB);
for(int i=0;i<dimBB;i++) Ep22[i]=d_vec[i];
for(int i=0;i<dimBB;i++) for(int j=0;j<dimBB;j++)
                                Vp22[i][j]=v_mat[i][j];

for(int i=0;i<dimBB;i++) delete [] h_mat[i];delete [] h_mat;
for(int i=0;i<dimBB;i++) delete [] v_mat[i];delete [] v_mat;
delete [] d_vec;
}
//===============================================================//
void mkHp11()
{
double tamp[dimBB][dimBB];
for(int i=0;i<dimBB;i++)
{
for(int j=0;j<dimBB;j++)
{
tamp[i][j]=0.0;
for(int k=0;k<dimBB;k++) tamp[i][j]+=H11[i][k]*Vp11[k][j];
}
}
for(int i=0;i<dimBB;i++)
{
for(int j=0;j<dimBB;j++)
{
Hp11[i][j]=0.0;
for(int k=0;k<dimBB;k++) Hp11[i][j]+=Vp11[k][i]*tamp[k][j];
}
}
double error=0.0;
for(int i=0;i<dimBB;i++)
{
for(int j=0;j<dimBB;j++)
{
if((fabs(Hp11[i][j]>error))&&(i!=j)) error=fabs(Hp11[i][j]);
}
}
printf("Hp11_error=%18.18f\n",error);
}
//===============================================================//
void mkHp22()
{
double tamp[dimBB][dimBB];
for(int i=0;i<dimBB;i++)
{
```

```cpp
for(int  j=0;j<dimBB; j++)
{
tamp[ i ][ j ]=0.0;
for(int  k=0;k<dimBB;k++)  tamp[ i ][ j ]+=H22[ i ][ k ]*Vp22[ k ][ j ];
}
}
for(int  i=0;i<dimBB; i++)
{
for(int  j=0;j<dimBB; j++)
{
Hp22[ i ][ j ]=0.0;
for(int  k=0;k<dimBB;k++)  Hp22[ i ][ j ]+=Vp22[ k ][ i ]*tamp[ k ][ j ];
}
}
double  error=0.0;
for(int  i=0;i<dimBB; i++)
{
for(int  j=0;j<dimBB; j++)
{
if((( fabs(Hp22[ i ][ j]>error ))&&(i !=j ))  error=fabs(Hp22[ i ][ j ]);
}
}
printf("Hp22_error=%18.18f\n", error );
}
//===============================================================//
void  mkHp12()
{
double  tamp[dimBB ][ dimBB ];
for(int  i=0;i<dimBB; i++)
{
for(int  j=0;j<dimBB; j++)
{
tamp[ i ][ j ]=0.0;
for(int  k=0;k<dimBB;k++)  tamp[ i ][ j ]+=H12[ i ][ k ]*Vp22[ k ][ j ];
}
}
for(int  i=0;i<dimBB; i++)
{
for(int  j=0;j<dimBB; j++)
{
Hp12[ i ][ j ]=0.0;
for(int  k=0;k<dimBB;k++)  Hp12[ i ][ j ]+=Vp11[ k ][ i ]*tamp[ k ][ j ];
}
}
}
//===============================================================//
void  mkHp21()
{
double  tamp[dimBB ][ dimBB ];
for(int  i=0;i<dimBB; i++)
{
for(int  j=0;j<dimBB; j++)
```

```cpp
{
tamp[i][j]=0.0;
for(int k=0;k<dimBB;k++) tamp[i][j]+=H21[i][k]*Vp11[k][j];
}
}
for(int i=0;i<dimBB;i++)
{
for(int j=0;j<dimBB;j++)
{
Hp21[i][j]=0.0;
for(int k=0;k<dimBB;k++) Hp21[i][j]+=Vp22[k][i]*tamp[k][j];
}
}
}
//=====================================================================//

void diagH()
{
double **h_mat;h_mat=new double *[dimH];
for(int i=0;i<dimH;i++) h_mat[i]=new double[dimH];
double **v_mat;v_mat=new double *[dimH];
for(int i=0;i<dimH;i++) v_mat[i]=new double[dimH];
double *d_vec=new double[dimH];

for(int i=0;i<dimH;i++) for(int j=0;j<dimH;j++) h_mat[i][j]=H[i][j];
jacobi(h_mat,dimH,d_vec,v_mat,10e-36);
eigsrt(d_vec,v_mat,dimH);
for(int i=0;i<dimH;i++) Ep[i]=d_vec[i];
for(int i=0;i<dimH;i++) V11[i]=v_mat[0][i]*v_mat[0][i];
for(int i=0;i<dimH;i++) V12[i]=v_mat[0][i]*v_mat[dimBB][i];

for(int i=0;i<dimH;i++) delete [] h_mat[i];delete [] h_mat;
for(int i=0;i<dimH;i++) delete [] v_mat[i];delete [] v_mat;
delete [] d_vec;
}
//=====================================================================//
void sumrule()
{
double sumrule11=0.0;
for(int i=0;i<dimH;i++) sumrule11+=V11[i];
double sumrule12=0.0;
for(int i=0;i<dimH;i++) sumrule12+=V12[i];
printf("sr11=%18.18f sr12=%18.18f\n",sumrule11,sumrule12);
double sumruless=0.0;
for(int i=0;i<dimH;i++) sumruless+=Vss[i];
double sumruleaa=0.0;
for(int i=0;i<dimH;i++) sumruleaa+=Vaa[i];
printf("srss=%18.18f sraa=%18.18f\n",sumruless,sumruleaa);

}
//=====================================================================//
void mkVsa()
```

```cpp
{
for(int  i=0;i<dimH;i++)  Vss[i]=V11[i]+V12[i];
for(int  i=0;i<dimH;i++)  Vaa[i]=V11[i]-V12[i];
}
//============================================================//
void saveEV()
{
FILE*fp;
fp=fopen("ev11.dat","w");
for(int  i=0;i<dimH;i++)
{
fprintf(fp,"%18.18f_%18.18f\n",Ep[i],V11[i]);
}
fclose(fp);
fp=fopen("ev12.dat","w");
for(int  i=0;i<dimH;i++)
{
fprintf(fp,"%18.18f_%18.18f\n",Ep[i],V12[i]);
}
fclose(fp);
fp=fopen("evSS.dat","w");
for(int  i=0;i<dimH;i++)
{
fprintf(fp,"%18.18f_%18.18f\n",Ep[i],Vss[i]);
}
fclose(fp);
fp=fopen("evAA.dat","w");
for(int  i=0;i<dimH;i++)
{
fprintf(fp,"%18.18f_%18.18f\n",Ep[i],Vaa[i]);
}
fclose(fp);
}
//============================================================//
main()
{
mkNB();
mkH11();mkH12();
mkH21();mkH22();
//seeH11();
diagH11();
diagH22();
mkHp11();
mkHp22();
mkHp12();
mkHp21();
mkH();
diagH();
mkVsa();
sumrule();
saveEV();
}
```

//==//

white.cc

```cpp
//================================================================//
// white . cc
//================================================================//
// Holstein  2  sites
// polaron from density  matrix
//================================================================//
#include <cmath>
#include <cstdio>
#include <cstdlib>
#include "jacobi .h"
#include "eigsrt .h"
//================================================================//
const double  e0 =0.0;
const double  w0=1.0;
const double  t0 =1.0;
const double  g0 =1.0;
const int  dimB =25;
const int  dimBB=dimB*dimB ;
const int  dimH=2*dimBB ;
//================================================================//
int  NB1[dimBB ] ,NB2[dimBB ] ,NB[dimBB ] ;
double  H11 [dimBB ] [dimBB ] ,H12 [dimBB ] [dimBB ] ;
double  H21 [dimBB ] [dimBB ] ,H22 [dimBB ] [dimBB ] ;
double  H[dimH ] [dimH ] ;
double  Ep [dimH ] ,Vp [dimH ] [dimH ] ;
double  R0[dimB ] [dimB ] ,RP[dimB ] [dimB ] ;
double  ER0[dimB ] ,ERP[dimB ] ;
double  U0[dimB ] [dimB ] ,UP[dimB ] [dimB ] ;
double  G11 [dimB ] [dimB ] [dimH ] ,G12 [dimB ] [dimB ] [dimH ] ;
double  P110 [dimH ] ,P120 [dimH ] ,P11P [dimH ] ,P12P [dimH ] ;
//================================================================//
void  mkNB ()
{
for (int  n1 =0;n1<dimB ; n1 ++)
{
for (int  n2 =0;n2<dimB ; n2 ++)
{
NB1[n1+dimB*n2 ]=n1 ;
NB2[n1+dimB*n2 ]=n2 ;
NB[n1+dimB*n2 ]=n1+n2 ;
}
}
}
//================================================================//
void  seeNB ()
{
for (int  i =0;i<dimBB ; i ++)
printf ("%d_n1=%d_n2=%d\n" , i ,NB1[i ] ,NB2[i ] );
```

```cpp
}
//=================================================================//
void mkH11()
{
for(int m=0;m<dimBB;m++) for(int n=0;n<dimBB;n++) H11[m][n]=0.0;
for(int n=0;n<dimBB;n++) H11[n][n]=e0+w0*NB[n];
for(int m=0;m<dimBB;m++)
{
for(int n=0;n<dimBB;n++)
{
int tutu=0;
if(NB2[m]==NB2[n]) tutu+=1;
if(NB1[m]==(NB1[n]-1)) tutu+=1;
if(tutu==2)
{
H11[m][n]=-g0*w0*sqrt(1.0*NB1[n]);
H11[n][m]=-g0*w0*sqrt(1.0*NB1[n]);
}
}
}
}
//=================================================================//
void mkH22()
{
for(int m=0;m<dimBB;m++) for(int n=0;n<dimBB;n++) H22[m][n]=0.0;
for(int n=0;n<dimBB;n++) H22[n][n]=e0+w0*NB[n];
for(int m=0;m<dimBB;m++)
{
for(int n=0;n<dimBB;n++)
{
int tutu=0;
if(NB1[m]==NB1[n]) tutu+=1;
if(NB2[m]==(NB2[n]-1)) tutu+=1;
if(tutu==2)
{
H22[m][n]=-g0*w0*sqrt(1.0*NB2[n]);
H22[n][m]=-g0*w0*sqrt(1.0*NB2[n]);
}
}
}
}
//=================================================================//
void mkH12()
{
for(int m=0;m<dimBB;m++) for(int n=0;n<dimBB;n++) H12[m][n]=0.0;
for(int m=0;m<dimBB;m++) H12[m][m]=-t0;
}
//=================================================================//
void mkH21()
{
for(int m=0;m<dimBB;m++) for(int n=0;n<dimBB;n++) H21[m][n]=0.0;
for(int m=0;m<dimBB;m++) H21[m][m]=-t0;
```

```cpp
}
//===============================================================//
void mkH()
{
for(int m=0;m<dimBB;m++)
{
for(int n=0;n<dimBB;n++)
{
H[m][n]=H11[m][n];
H[m][n+dimBB]=H12[m][n];
H[m+dimBB][n]=H21[m][n];
H[m+dimBB][n+dimBB]=H22[m][n];
}
}
}
//===============================================================//
void seeH11()
{
printf("H11:\n");
for(int i=0;i<dimBB;i++)
{
for(int j=0;j<dimBB;j++) printf("%2.2f_",H11[i][j]);
printf("\n");
}
}
//===============================================================//
void seeH22()
{
printf("H22:\n");
for(int i=0;i<dimBB;i++)
{
for(int j=0;j<dimBB;j++) printf("%2.2f_",H22[i][j]);
printf("\n");
}
}
//===============================================================//
void seeH()
{
for(int i=0;i<dimH;i++)
{
for(int j=0;j<dimH;j++) printf("%2.2f_",H[i][j]);
printf("\n");
}
}
//===============================================================//
void diagH()
{
double **h_mat;h_mat=new double *[dimH];
for(int i=0;i<dimH;i++) h_mat[i]=new double[dimH];
double **v_mat;v_mat=new double *[dimH];
for(int i=0;i<dimH;i++) v_mat[i]=new double[dimH];
double *d_vec=new double[dimH];
```

```cpp
for(int i=0;i<dimH;i++) for(int j=0;j<dimH;j++) h_mat[i][j]=H[i][j];
jacobi(h_mat,dimH,d_vec,v_mat,10e-24);
eigsrt(d_vec,v_mat,dimH);
for(int i=0;i<dimH;i++) Ep[i]=d_vec[i];
for(int i=0;i<dimH;i++) for(int j=0;j<dimH;j++)
                        Vp[i][j]=v_mat[i][j];

for(int i=0;i<dimH;i++) delete [] h_mat[i]; delete [] h_mat;
for(int i=0;i<dimH;i++) delete [] v_mat[i]; delete [] v_mat;
delete [] d_vec;
}
//=================================================================//
void mkR0()
{
for(int i=0;i<dimB;i++) for(int j=0;j<dimB;j++) R0[i][j]=0.0;
for(int i=0;i<dimB;i++)
{
for(int j=0;j<dimB;j++)
{
for(int m2=0;m2<dimB;m2++)
{
int posi=i+m2*dimB;
int posj=j+m2*dimB;
R0[i][j]+=Vp[posi][0]*Vp[posj][0];
}
}
}
}
//=================================================================//
void mkRP()
{
for(int i=0;i<dimB;i++) for(int j=0;j<dimB;j++) RP[i][j]=0.0;
for(int i=0;i<dimB;i++)
{
for(int j=0;j<dimB;j++)
{
for(int m2=0;m2<dimB;m2++)
{
int posi=i+m2*dimB;
int posj=j+m2*dimB;
RP[i][j]+=Vp[posi][1]*Vp[posj][1];
}
}
}
}
//=================================================================//
void diagR0()
{
double **h_mat;h_mat=new double *[dimB];
for(int i=0;i<dimB;i++) h_mat[i]=new double[dimB];
double **v_mat;v_mat=new double *[dimB];
```

```cpp
  for(int  i=0;i<dimB ; i++)  v_mat [ i ]=new  double [dimB ];
  double *d_vec=new  double [dimB ];

  for(int  i=0;i<dimB ; i++)  for(int  j=0;j<dimB ; j++)
                              h_mat [ i ][ j ]=R0[ i ][ j ];
  jacobi (h_mat ,dimB , d_vec , v_mat ,10 e −24);
  eigsrt ( d_vec , v_mat , dimB );

  for(int  i=0;i<dimB ; i++)  ER0[ i ]=d_vec [dimB−1−i ];
  for(int  i=0;i<dimB ; i++)  for(int  j=0;j<dimB ; j++)
                              U0[ i ][ j ]=v_mat [ i ][ dimB−1−j ];

  for(int  i=0;i<dimB ; i++)  delete [] h_mat [ i ]; delete [] h_mat ;
  for(int  i=0;i<dimB ; i++)  delete [] v_mat [ i ]; delete [] v_mat ;
  delete [] d_vec ;
}
//=================================================================//
void  diagRP ()
{
double **h_mat ; h_mat=new  double  *[dimB ];
for(int  i=0;i<dimB ; i++)  h_mat [ i ]=new  double [dimB ];
double **v_mat ; v_mat=new  double  *[dimB ];
for(int  i=0;i<dimB ; i++)  v_mat [ i ]=new  double [dimB ];
double *d_vec=new  double [dimB ];

  for(int  i=0;i<dimB ; i++)  for(int  j=0;j<dimB ; j++)
                              h_mat [ i ][ j ]=RP[ i ][ j ];
  jacobi (h_mat ,dimB , d_vec , v_mat ,10 e −24);
  eigsrt ( d_vec , v_mat , dimB );

  for(int  i=0;i<dimB ; i++)  ERP[ i ]=d_vec [dimB−1−i ];
  for(int  i=0;i<dimB ; i++)  for(int  j=0;j<dimB ; j++)
                              UP[ i ][ j ]=v_mat [ i ][ dimB−1−j ];


  for(int  i=0;i<dimB ; i++)  delete [] h_mat [ i ]; delete [] h_mat ;
  for(int  i=0;i<dimB ; i++)  delete [] v_mat [ i ]; delete [] v_mat ;
  delete [] d_vec ;
}
//=================================================================//
void  mkG11 ()
{
for(int  m=0;m<dimB ;m++)
{
for(int  n=0;n<dimB ; n++)
{
for(int  p=0;p<dimH ; p++)
{
G11 [m][ n ][ p ]=Vp[m][ p ]*Vp[ n ][ p ];
}
}
}
```

```cpp
}
//===============================================================//
void mkG12()
{
for(int m=0;m<dimB;m++)
{
for(int n=0;n<dimB;n++)
{
for(int p=0;p<dimH;p++)
{
G12[m][n][p]=Vp[m][p]*Vp[dimBB+dimB*n][p];
}
}
}
}
//===============================================================//
void saveG11()
{
FILE *fp;
fp=fopen("G11.dat","w");
for(int p=0;p<dimH;p++)
fprintf(fp,"%18.18f_%18.18f\n",Ep[p],G11[0][0][p]);
fclose(fp);
}
//===============================================================//
void saveG12()
{
FILE *fp;
fp=fopen("G12.dat","w");
for(int p=0;p<dimH;p++)
fprintf(fp,"%18.18f_%18.18f\n",Ep[p],G12[0][0][p]);
fclose(fp);
}
//===============================================================//
void saveP110()
{
FILE *fp;
fp=fopen("P110.dat","w");
for(int p=0;p<dimH;p++)
fprintf(fp,"%18.18f_%18.18f\n",Ep[p],P110[p]);
fclose(fp);
}
//===============================================================//
void saveP120()
{
FILE *fp;
fp=fopen("P120.dat","w");
for(int p=0;p<dimH;p++)
fprintf(fp,"%18.18f_%18.18f\n",Ep[p],P120[p]);
fclose(fp);
}
//===============================================================//
```

```cpp
void saveP11P()
{
FILE *fp;
fp=fopen("P11P.dat","w");
for(int p=0;p<dimH;p++)
fprintf(fp,"%18.18f_%18.18f\n",Ep[p],P11P[p]);
fclose(fp);
}
//==================================================================//
void saveP12P()
{
FILE *fp;
fp=fopen("P12P.dat","w");
for(int p=0;p<dimH;p++)
fprintf(fp,"%18.18f_%18.18f\n",Ep[p],P12P[p]);
fclose(fp);
}
//==================================================================//
void mkP110()
{
for(int p=0;p<dimH;p++)
{
P110[p]=0.0;
for(int i=0;i<dimB;i++)
{
for(int j=0;j<dimB;j++)
{
P110[p]+=U0[i][0]*G11[i][j][p]*U0[j][0];
}
}
}
}
//==================================================================//
void mkP120()
{
for(int p=0;p<dimH;p++)
{
P120[p]=0.0;
for(int i=0;i<dimB;i++)
{
for(int j=0;j<dimB;j++)
{
P120[p]+=U0[i][0]*G12[i][j][p]*U0[j][0];
}
}
}
}
//==================================================================//
void mkP11P()
{
for(int p=0;p<dimH;p++)
{
```

```cpp
P11P[p]=0.0;
for(int  i =0;i<dimB; i ++)
{
for(int  j =0;j<dimB; j ++)
{
P11P[p]+=UP[ i ][0]*G11[ i ][ j ][p]*UP[ j ][0];
}
}
}
}
//========================================================================//
void  mkP12P ()
{
for(int  p=0;p<dimH; p++)
{
P12P[ p ]=0.0;
for(int  i =0;i<dimB; i ++)
{
for(int  j =0;j<dimB; j ++)
{
P12P[p]+=UP[ i ][0]*G12[ i ][ j ][p]*UP[ j ][0];
}
}
}
}
//========================================================================//
void  sumrule ()
{
double  sr110 =0.0;
double  sr120 =0.0;
double  sr11P =0.0;
double  sr12P =0.0;
for(int  p=0;p<dimH; p++)
{
sr110+=P110[p ];
sr120+=P120[p ];
sr11P+=P11P[p ];
sr12P+=P12P[p ];
}
printf("sr110=%18.18f\n",sr110 );
printf("sr120=%18.18f\n",sr120 );
printf("sr11P=%18.18f\n",sr11P );
printf("sr12P=%18.18f\n",sr12P );
}
//========================================================================//
main ()
{
mkNB ();
//seeNB ();
mkH11 ();mkH12 ();
mkH21 ();mkH22 ();
//seeH11 ();
```

```cpp
// seeH22 ( );
mkH ( );
diagH ( );
printf ( "E0=%18.18f\n" , Ep [ 0 ] );
printf ( "E1=%18.18f\n" , Ep [ 1 ] );
mkR0 ( );
mkRP ( );
diagR0 ( );
diagRP ( );
for ( int  p=0;p<dimB ; p++ )  printf ( "R0_%d_%lf \n" , p , ER0[ p ] );
for ( int  p=0;p<dimB ; p++ )  printf ( "RP_%d_%lf \n" , p , ERP[ p ] );
mkG11 ( );
mkG12 ( );
saveG11 ( );
saveG12 ( );
mkP110 ( );
mkP120 ( );
saveP110 ( );
saveP120 ( );
mkP11P ( );
mkP12P ( );
saveP11P ( );
saveP12P ( );
sumrule ( );
}
//===============================================================//
```

Made in the USA
Monee, IL
07 July 2026